ANALYTICAL TROUBLESHOOTING OF PROCESS MACHINERY AND PRESSURE VESSELS

ANALYTICAL TROUBLESHOOTING OF PROCESS MACHINERY AND PRESSURE VESSELS

Including Real-World Case Studies

ANTHONY SOFRONAS
Kingwood, Texas

WILEY-INTERSCIENCE

A JOHN WILEY & SONS, INC., PUBLICATION

Library of Congress Cataloging-in-Publication Data:

Sofronas, Anthony.
  Analytical troubleshooting of process machinery and pressure vessels:
including real-world case studies / Anthony Sofronas.
      p. cm.
  Includes index.
  ISBN-13: 978-0-471-73211-2 (cloth)
  ISBN-10: 0-471-73211-7 (cloth)
1. Machinery—Maintenance and repair. 2. Plant maintenance. I. Title.
  TJ153.S6375 2005
  621.8'16—dc22   2005009734

Printed in the United States of America

10 9 8 7 6 5 4 3 2 1

To My Family
Cruz, Steve, Maria

CONTENTS

PREFACE

Purpose of the Book

A book such as this can only be written after one has worked in industry for many years. The case histories in this book represent a collection of over 40 years of problem solving by the author. The examples are from several industries I have worked in, including the petrochemical, transportation, and component manufacturing industries.

Early in my career I realized that time was not available to become an expert in all of the pieces of equipment that I had to design or troubleshoot. Often, a piece of equipment was the only one of its kind, such as a brake manufacturing grinding tool or a special drier or extruder. By analyzing the equipment using simple mathematics in analytical models, the models could be used to obtain a better understanding of the equipment operation and failure modes. The models allowed me to examine the internal workings of the equipment, to talk intelligibly with manufacturers, to understand changes that needed to be made for continued operation, and to grow technically in my career. My desire is to transfer this knowledge to others with the help of this book. The material in the book has been presented in engineering seminars for many years and thus has input from many participants, and as such represents an enhancement of the course notes.

This is not a book about complex analytical solutions. It is a practical book that has been used to design and troubleshoot over 90% of the equipment worked. A rough estimate is that the examples in this book have saved over $50 million in lost production or warranty claims by eliminating repeat failures or by avoiding failures altogether. This is because the failure cause had been well understood and prevented from occurring or recurring. These savings included the efforts of many other talented specialists who worked the details involved in implementing the solutions, which is usually the most difficult part. Attention to the details can make the difference between a successful startup and a painful one.

High horsepower, speeds, temperatures, pressures, and abnormal operating conditions has made equipment engineering so complex that intuition and past experiences cannot be depended upon solely to solve problems. Safety, production loss concerns, and environmental and legal issues can be so great for a company that decisions need to be backed with sound troubleshooting techniques and documentation. It is the goal of this book to help the reader to learn some of these techniques.

Content and Arrangement

Chapters 1 and 2 are a review of the engineering tools used in analyzing machinery and pressure vessel components. Actual case histories are used to illustrate the principles learned. Loads, stresses, deflections, and components such as bearings, gears, bolts, and gaskets are explained. Wear is also reviewed, and equations for determining the wear life are shown by example.

Chapter 3 deals with the vibration of machines and structures. Simplified solutions for linear and torsional machinery vibrations are provided, as are case histories of machinery, heat exchangers, and piping vibration problems.

Chapter 4 is concerned with fluid flow. After a review of the basics of fluid flow and energy equations, the sizing of centrifugal pumps is examined in detail. Forces of fluids and pressure-loss equations are shown and used to solve problems from racing car stability to piping losses.

In Chapter 5 we review heat transfer and conduction, convection, and radiation. Heat exchanger sizing case histories and the temperature determination to troubleshoot bearings, resistors, and piping are presented.

Chapter 6 covers compressor systems and the thermodynamics principles used in these systems. Gas laws and energy equations are used to analyze centrifugal and reciprocating compressors, and troubleshooting case histories are included. Mechanical seals and couplings are explained, as they represent failure areas associated with compressor and pumping equipment.

Chapter 7 takes us into the area of statistics and reliability and their use in troubleshooting. Case histories are used to explain how data can be used in troubleshooting.

In Chapter 8 we utilize a unique method to organize data and information so that an orderly decision can be made on the evidence presented. By case histories, methods are shown on how to use the troubleshooting methods of previous chapters in an organized manner. There is usually not a single cause that is responsible for a failure, and this chapter helps to identify and address less obvious causes.

In Chapter 9 we examine materials used in the construction of machines and pressure vessels. Steels and their properties are reviewed, as are the failure modes of shafts, bolting, structures, and vessels. Fretting corrosion is discussed in detail along with case histories because of its seriousness in causing fatigue failures.

Chapters 10 and 11 contain case histories on machinery and pressure vessel failures. The analysis methods and problem-solving techniques presented in previous chapters are applied to troubleshooting these potentially high-impact failure areas.

A list of references used in preparing the book follows the final chapter.

Acknowledgments

I wish to acknowledge Richard Gill, my colleague and friend, for his many helpful suggestions throughout the years.

Dr. Khalil Taraman and Dr. William Spurgeon, my doctorate advisors and friends, have greatly influenced my career by their works and work ethics.

Heinz Bloch, with his guidance and friendship, together with his prolific book-writing ability, was an inspiration to me to delve into this book-writing endeavor.

I also wish to thank *Hydrocarbon Processing* magazine, especially Les Kanes, for publishing many of the case histories throughout the years.

Many thanks also go to John Wiley & Sons, Inc., especially Bob Esposito and Jonathan Rose, for agreeing to publish this work and for their many valuable suggestions.

Over the years, Charlie Arnold, Rich Skinner, Arlon Hokett, Don Holy, Geoff Kinison, and Marty Hapeman have provided valuable contributions to my machinery knowledge, for which I thank them.

All of this couldn't have been accomplished without the help of my wife, Mrs. Cruz Velasquez Sofronas. She provided careful typing and preliminary editing of the manuscript along with continual encouragement to produce this work. I thank her for always being there for me.

ANTHONY SOFRONAS

1

INTRODUCTION

The author has had the privilege of working with both machinery and fixed equipment problems during his 40-year career in the transportation and hydrocarbon processing industries. Machines certainly have a personality of their own, but fixed equipment, when it fails, can result in catastrophic and lengthy downtime delays. Any failure must be understood and remedied so that it cannot reoccur. The case histories in this book were used to solve actual problems experienced by the author. Some were used to simplify a problem so that it could be explained to others in group failure analysis team efforts.

These case histories are not trivial and represent value added of over $50 million. The majority of the losses avoided are not maintenance related but rather, avoid business losses due to an unplanned shutdown of an operating plant for an extended period of time. Some of the case histories prevented failures from occurring or reduced downtime when they did occur, and others avoided repeat failures.

This book is unique in that it doesn't just present equations but also illustrates how to simplify complex systems so that the formulas can be applied. Most of the analyses are not exact but are important since they provide a much more detailed understanding of the problem and were enough to implement successful solutions in industry. Exact numerical values are not always required for the types of problems addressed in this book. When a stress is calculated to be $10,000\,lb/in^2$ and the stress required to fail is $60,000\,lb/in^2$, it is obvious that more extensive calculations are not warranted and that a decision can be made. In other cases the calculations may be so close to the failure stress and of such a critical nature that further detailed analysis is required.

Analytical Troubleshooting of Process Machinery and Pressure Vessels: Including Real-World Case Studies, by Anthony Sofronas
Copyright © 2006 John Wiley & Sons, Inc.

An engineer may have spent one day on an analysis to recommend that \$50,000 be invested for a finite-element study by an outside firm.

Chapters 2 through 9 are designed to illustrate the basic equations and rules of thumb that engineers should know and show how basic engineering methods can be used to help solve complex equipment problems. In Chapters 10 and 11 we utilize these equations and more to model many actual problems experienced by the author. They deal with analytical troubleshooting, one of the most cost-effective efforts an engineer can perform. It is the difference between speculating as to the cause of a failure or determining and validating the cause. It is the difference between making decisions and testing them in the field under production conditions or testing them in the office on a computer or on paper. A failure in the first case requires shutdown and repair of the equipment. A failure on paper requires a new piece of paper for another "test." This is what engineering is about and why engineers are employed. An engineer's job is to analyze and solve, not to change parts.

Analytical modeling of systems means representing the performance of equipment with sketches and equations in order to predict the operation of the equipment. There is a reason that you will see simple sketches rather than elaborate drawings in this book. In basic modeling, such sketches are usually all that is needed for a fundamental understanding of a problem. The sketches can always be refined later.

The concept of analytical modeling in engineering can be quite intimidating to an engineer requiring solutions to actual field problems. It brings to mind college, calculus, and developing and solving differential or integral equations. As many practicing engineers have found out, engineering in our profession isn't really like this. Calculus was used by others more gifted in the field of mathematics; simple algebraic expressions have been developed that we use to solve most of our engineering problems. We can go through most of our careers relying only on algebra, trigonometry, and geometry and be very successful problem solvers and machine builders. These models can be as simple or as complex as is necessary to solve the problem. In this book, things are kept simple.

Most of the simplified solutions used in this book were backed up with more elaborate solutions. More accurate results were used when the consequence of a failure was unacceptable. The finite-element method may have been performed in addition to a simple stress, heat transfer, or vibration analysis. One sacrifice that is made with a more complex solution is that much of the understanding that comes from a simple analysis is lost. A three-dimensional piping isometric of the piping of a process may be quite elaborate and confusing when considering the thermal loads on a vessel. Suppose that the problem can be reduced to one large pipe to a vessel, which has all the temperature change. All the other 20 lines may be of small diameter and experience no temperature effect. A very simple and understandable model then evolves. The result can be explained easily to most anyone. It will also allow a check on the more complex system results, if one is required. The more complex, the more chance for computational errors. A simple order-of-magnitude check from a simple analysis is always welcome.

The real key to simplifying complex equipment is in understanding the operation of the equipment. Only when equipment is understood can it be simplified. In the piping example the person modeling the system has experience which says that the loads on the vessel are not influenced much by the long small-diameter pipes, as they are very flexible. The ability to visualize and simplify systems is more of an art than a science but fortunately, something most engineers are capable of mastering.

The goal of this book is to show the equipment specialist that by reducing an equipment system to an analytical model, the model can be used to identify the cause of a failure. It does this by allowing the engineer to better understand what is occurring and to replace opinions with quantitative results showing what is occurring and what effect modifications will make.

We illustrate how important it is to know the mode of failure, but also that knowing the mode alone may not solve a problem. An understanding and control of the variables by analysis is necessary to keep failures from reoccurring. It is our intent in the book to provide additional tools necessary to strengthen a specialist's technical contribution to industry and to his or her career.

2

STRENGTH OF MATERIALS

2.1 LOAD CALCULATIONS

In stationary and dynamic structures and machinery, in many cases determination of the loads imposed defines the adequacy of the equipment. Forces and moments on a structure are important to define, as they are needed to define stresses, deflections, bearing life, and so on. Two primary equilibrium equations are force equilibrium and moment equilibrium.

For *force equilibrium*, the sum of the forces must equal zero for a stationary unit:

$$+ \downarrow \Sigma F = 0$$

For *moment equilibrium*, the sum of the moments must equal zero for a stationary unit:

$$\text{ccw} + \Sigma M = 0$$

Here ccw represents counterclockwise rotation. To illustrate the use of these equations, consider the loading of a large motor driving a gear unit via a multiple V-belt (Figure 2.1). All that is important here is that the belt exerts a force F_3 on the shaft and that the bearing loads F_1 and F_2 must be found. The rotor weight W was not considered in this example.

Analytical Troubleshooting of Process Machinery and Pressure Vessels: Including Real-World Case Studies, by Anthony Sofronas

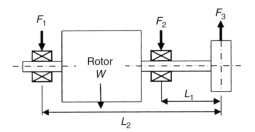

FIGURE 2.1 Free-body diagram belt-driven motor.

The only known value is that of F_3, which can be determined from the horse-power transmitted. There are two unknowns; however, since there are also two equations, the problem is statically determinant.

$$+ \downarrow \Sigma F = 0$$

$$F_1 + F_2 - F_3 = 0$$

Setting counterclockwise rotation as + and summing the moments about F_2 gives us

$$\text{ccw} + \Sigma M = 0$$

$$F_3 L_1 + F_1 (L_2 - L_1) = 0$$

$$F_1 = -\frac{F_3 L_1}{L_2 - L_1}$$

Note how summing about F_2 eliminated this unknown since there was no moment arm. Using sum of forces yields

$$F_2 = F_3 - F_1$$

Torque, which is a twisting moment on a shaft, is handled in the same manner as moments:

$$\Sigma T = 0$$

This is important when evaluating shaft stresses, and the similarities will be obvious when discussing stresses.

With the belt loads calculated and the bearing loads known, we could now proceed to determine the bearing life and shaft stresses. This brings us to the next subject.

2.2 STRESS CALCULATIONS

Valuable test data on engineering materials such as metals are tabulated as stress-to-failure information. This is the load in pounds that material can take per square inch of area. Tensile stress tends to elongate a longitudinal fiber and usually is represented by a plus sign; compressive stresses shorten a fiber and are represented by a minus sign.

Various types of loading result in different types of stresses. All or some may act on a part at the same time.

- Direct stress (axial stress)
- Shear stress
- Bending stress
- Torsional stress
- Combined stresses

As will be shown, each stress can be calculated independently and the stresses summed once the stress distribution is understood.

Figure 2.2 illustrates the case of several loading types being applied to a member. The diameters of the pin (d_p) and rod (d_r) are shown, as they enter into the stress calculations. In the figure, P is the axial load (lb), T the torque (in.-lb), and M the bending moment (in.-lb).

2.2.1 Axial Stress

The stress due to the axial loading (P) on the member in Figure 2.2 is axial stress:

$$S_{\text{axial}} = \frac{KP}{A} \qquad \text{lb/in}^2$$

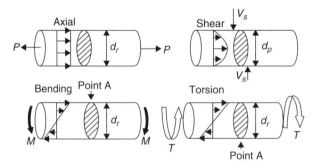

FIGURE 2.2   Axial, shear, bending, and torsional stress.

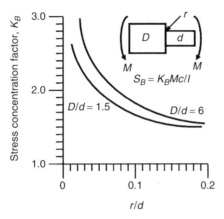

FIGURE 2.3 Bending stress concentration factor.

where

$$A_{\text{solid rod}} = \frac{\pi}{4}d_r^2$$

$$A_{\text{hollow rod}} = \frac{\pi}{4}(d_{r\,\text{outside}}^2 - d_{r\,\text{inside}}^2)$$

$$A_{\text{rectangle}} = \text{base} \times \text{width}$$

In this case, since a fiber is being elongated, it represents a positive sign or tensile stress. A fiber being shortened would be in compression and would carry a negative sign. For the case shown, the stress is evenly distributed over the surface, so the stress concentration factor K is 1. If there were a stress increaser such as a stepped shaft, K might be greater than 1.

Graphs, charts, and tables are available for determining K in most cases [1]. Figure 2.3 represents a stepped shaft under bending loading. Torsional loading is very similar. Notice how a sharp radius affects the stress level. The prudent designer would keep r/d greater than 0.1 if possible. This would keep the stress concentration factor K low. When one talks nominal stress, one is essentially talking of no stress concentration factor ($K = 1$).

2.2.2 Shear Stress

Shear stress is being experienced on diameter d_p in Figure 2.2, thus producing shearing stresses. The shear loads are V_s and the shear stress is

$$S_{\text{shear}} = \frac{K_1 V_s}{A_s} \qquad \text{lb/in}^2$$

$$A_s = \frac{\pi}{4}d_p^2 \qquad \text{in}^2$$

For shear of circular sections $K_1 = \frac{4}{3}$, and for square sections, $K_1 = \frac{3}{2}$.

The important fact is that of the stress distribution. The direct shear stress is maximum at the center and zero at the surface for shafts. This becomes important when adding shearing and bending stresses. Within limits, the shearing stresses have little effect on the magnitude of bending stresses.

2.2.3 Bending Stress

Bending stress is shown on diameter d_r in Figure 2.2. The controlling equation for bending stress is

$$S_B = \frac{K_B M c}{I} \quad \text{lb/in}^2$$

where c is the distance from the neutral axis to the stress point desired, and for now is considered to be the outer surface. For the shaft d_r shown, this is simply

$$c = \frac{d_r}{2} \quad \text{in.}$$

I is the area moment of inertia and is the cross section that resists bending. For a solid shaft,

$$I = \frac{\pi}{64} d_r^4 \quad \text{in}^4$$

For a hollow shaft,

$$I = \frac{\pi}{64} (d_{outside}^4 - d_{inside}^4) \quad \text{in}^4$$

For a rectangle,

$$I = \frac{bh^3}{12} \quad \text{in}^4$$

Values for other sections are tabulated in reference books.
  A useful relation for a solid shaft is

$$S_B = \frac{32 K M}{\pi d_r^3} \quad \text{lb/in}^2$$

The stress distribution produced on the diameter d_r is due to moment M. The stress is maximum at A tension and 180° maximum as compression, and zero at the neutral axis. Also notice that this is quite different from the shear case. The shear was maximum at the neutral axis and zero at the outer surface. The stress direction is parallel to the plane of the paper or, stated another way, in the same direction as the bending moment.

2.2.4 Torsional Stress

Torsional loading, or the shaft twisting moment, results in the torsional shearing stresses shown in Figure 2.2. The controlling equation for torsional stress is

$$S_s = \frac{KTc}{J} \qquad \text{lb/in}^2$$

The equation is very similar to that of bending, but the direction is 90° away. In other words, the stress direction is circumferential rather than axial (recall the way the fibers are distorted).

Again, c is the distance from the neutral axis. For maximum stress at the surface,

$$c = \frac{d_r}{2} \qquad \text{in.}$$

J is similar to I but is called the *polar moment of inertia* and is the cross section that resists torsion. For a solid shaft,

$$J = \frac{\pi d_r^4}{32} \qquad \text{in}^4$$

For a hollow shaft,

$$J = \frac{\pi (d_{\text{outside}}^4 - d_{\text{inside}}^4)}{32} \qquad \text{in}^4$$

Other cross sections are more difficult to define, as they are based on experimental data.

A useful relation for torsional stress in a solid shaft due to torque T is

$$S_s = \frac{16KT}{\pi d_r^3} \qquad \text{lb/in}^2$$

This stress occurs around the outside diameter. The stress distribution is such that in torsion it is maximum at the surface and zero at the center.

Since the bending and torsional stresses are represented by similar equations, common observations can be made. Notice that for a hollow shaft, the resisting areas are a function of the outside and inside diameters raised to the fourth power. Using hollow shafting has little effect on stress for ratios of the inside diameter to the outside diameter below 0.5. However, weight and cost can be reduced significantly without affecting the strength.

2.2.5 Combined Stresses

Combining the stresses that have been discussed takes special consideration. As mentioned, torsional and bending stresses cannot just be added together, as they are oriented vectorially 90° apart.

For the uniaxial case, the following equivalent stress is useful for combining bending and torsion:

$$S_{\text{equivalent}} = (S_b^2 + 3S_s^2)^{1/2} \qquad \text{lb/in}^2$$

Axial stress is added or subtracted from the bending stress depending on the sign:

$$S_{\text{equivalent}} = [(S_b + S_a)^2 + 3S_s^2]^{1/2} \qquad \text{lb/in}^2$$

Because direct shear is zero where torsion and bending stresses are maximum (at the surface) and maximum where torsion and bending are zero (at the center), they are usually not combined.

2.2.6 Thermal Stresses

Thermal stresses are produced in a structure such as a restrained bar whenever the expansion or contraction that would normally occur due to a temperature change in a bar or a plate is restricted in some way. For example, if the ends of a bar are fixed and the bar is heated, thermal stresses will be produced in the bar. Thermal stresses can become very important in industries such as refineries and petrochemical plants, where vessels are cycled from cold to hot conditions many times per day. In these cases low cycle fatigue can be induced due to the high thermal stresses. In most cases critical designs are verified by analysis techniques such as the finite-element method (FEM); however, it is usually desirable to have some idea of the magnitude of the stresses.

Several cases are presented that have proven useful. In all the equations the material is steel and T is the temperature differential between the material and ambient in °F; Poisson's ratio, μ, is 0.3 for steel. For material other than steel, use the appropriate μ value and scale the 200 factor, which is the product of the modulus of elasticity and the coefficient of thermal expansion, by the new values.

Axial stress in a straight uniform bar restrained at the ends:

$$S = 200T \qquad \text{lb/in}^2$$

Stress in mth bar of parallel bars of equal length but different temperatures:

$$S_m = 200 \left(\frac{(T_1 + T_2 + T_3 + \cdots)}{n} - T_m \right) \qquad \text{lb/in}^2$$

For example, suppose that the stresses in a fixed-tube heat exchanger were desired and the design was such that a large amount of heat transfer was found to occur in the first tube (Figure 2.4). The stress in this tube, with $T_1 = 400°F$, is

$$S_1 = 200 \left(\frac{(400 + 200 + 200)}{3} - 400 \right)$$

$$= 27 \text{ ksi compression}$$

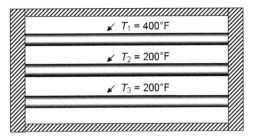

FIGURE 2.4   Parallel bar thermal stresses.

Stress throughout a flat plate fixed at all edges:

$$S = \frac{200T}{1-\mu}$$

Stress in a flat plate due to a thermal shock: If a plate or a vessel wall were suddenly subjected to temperature T, the maximum thermal shock stress that occurs at the surface of the plate or vessel is

$$S = \frac{200T}{1-\mu}$$

The stresses are compressive if heating is involved, and tensile with cooling. It should be mentioned that cracks can develop on the surface with rapid cooling. Internal cracks could develop with rapid heating since tensile stresses will be developed internally to balance the compressive stresses on the surface.

Transient thermal stresses of complex parts are difficult to analyze analytically, and finite-element methods are generally used on critical components. An example would be the analysis of the effect of a rapid heat-up rate on a pressure vessel. FEM depends on the film coefficients, which determine how fast or slow the heat is transferred into the component being analyzed. With small film coefficients the heat transfer rate can be much slower, and thus the thermal stresses much lower than those calculated from the simplified equations. High stresses calculated by approximate equations only indicate that a more detailed analysis may be necessary.

In all of the thermal stress equations above, buckling could occur before the stresses are reached. If buckling occurs first, it will usually limit the load and stress to a value less than the value of thermal stress.

2.2.7   Transient Temperatures and Stresses

In this section, we have discussed thermal stresses applied suddenly. This occurs when a metal surface is subjected quickly to the environmental temperature. The surface stress is approximated as

$$S = \begin{cases} 200\Delta T \text{ steel} & \text{lb/in}^2 \\ 130\Delta T \text{ aluminum} & \text{lb/in}^2 \end{cases}$$

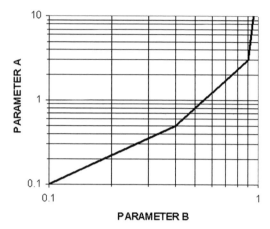

FIGURE 2.5   Parameters A and B.

Here ΔT is the difference between the applied temperature and the initial temperature on the metal surface. Often, the temperature is applied through a heat transfer
coefficient over a short time period, and the temperature can be less severe. This is
because the heat has to pass through a resistance path, the heat transfer coefficient h,
to reach the metal surface.

The method described next determines the change in surface temperature as a
function of the time, heat transfer coefficient, thermal conductivity k, and thermal
diffusivity α. A surface such as a plate is exposed to a temperature environment that
is applied to the heat transfer coefficient h for a period of time t.

In Figure 2.5 parameters A and B are defined as follows, with typical values given
for some of the constants:

$$A = (h/k)(\alpha t)^{1/2} \qquad\qquad \Delta T = B(T_{environment} - T_{initial})$$
$$\alpha_{steel} = 0.4 \text{ ft}^2/\text{hr} \qquad\qquad \alpha_{aluminum} = 3.6 \text{ ft}^2/\text{hr}$$
$$k_{steel} = 20 \text{ Btu/hr-ft-}°\text{F} \qquad k_{aluminum} = 100 \text{ Btu/hr-ft-}°\text{F}$$

The h values are given in Chapter 5. Water-cooled surface condensers with flows of
3 to 8 ft/sec have h values of 500 to 800 Btu/hr-ft^2-°F. Cryogenic fluids behave similarly to conventional fluids. When droplets are present, as in low-quality steam, the
value of h can increase by a factor of 10 as the droplets strike the surface.

In the method, parameter A is calculated from the data supplied, then the graph is
intercepted to find parameter B. With this value of B, use the following equation to
solve for ΔT:

$$\Delta T = B(T_{environment} - T_{initial})$$

This ΔT value can then be used in the stress equation. The following example illustrates this.

Calculate the stress on the surface of a steel block that is initially at 450°F and is subjected to an environment of 150°F for 0.01 hr. Use $k = 20$, $h = 200$, and $\alpha = 0.4$. Parameter A calculates as 0.632, and parameter B from the graph is 0.45.

$$\Delta T = B(T_{environment} - T_{initial})$$
$$= 0.45(150 - 450)$$
$$= -135°F$$

The stress on the surface is a compressive stress:

$$S = 200\Delta T \text{ steel}$$
$$= 200(-135) = -27,000 \text{ lb/in}^2$$

This is considerably lower than the surface stress developed by a suddenly applied temperature difference of 300°F, which is $-60,000$ lb/in² and for many steels would exceed the yield. Thus, short-duration applications of temperatures with small heat transfer coefficients will reduce the surface stresses developed. If it were desired to know the surface temperature of the metal, the following equation would be used:

$$T_{surface} = T_{initial} + \Delta T$$

2.2.8 High-Temperature Creep

One additional life-limiting failure mechanism is creep. Figure 2.6 shows a support bar in a furnace under a constant load W and a constant temperature T. Over time, with this constant stress, the bar will continue to elongate δ_{creep} with no additional load added. At some point in the future, this additional stretching will result in enough strain increase to cause the bar to rupture.

Various methods are used to predict the design life of structures under time, temperature, and stress conditions. For our cases we use the techniques shown in

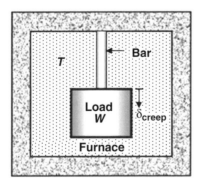

FIGURE 2.6 High-temperature creep.

TABLE 2.1   Lawson–Miller Parameters

| Minimum Rupture | LM Parameter | |
Stress (ksi)	Carbon Steel	316 Stainless Steel
30	—	29.2
25	—	29.8
20	28.2	30.6
15	30.0	31.6
10	31.8	33.0

Refs. [2] and [3]. This method utilizes a parameter that relates time and temperature to the long-term rupture strength of various materials. Only a brief review is given here, but complete curves are provided in the references cited.

The *Lawson–Miller* (LM) *parameter* is defined as

$$LM = (T + 460)(C + \log L_r) \times 10^{-3}$$

Here T is the bar temperature (°F), C is a constant (20 for ferritic and 15 for austenitic steels), and L_r is the rupture design life of the bar (hr). Some values of the Lawson–Miller parameter are given in Table 2.1.

The necessary steps are to insert the LM value and temperature and solve for L_r:

$$\log_{10} L_r = N$$
$$L_r = 10^N \qquad \text{hr}$$

The reader can proceed through the calculations for 316 stainless steel when the temperature of the bar is 1100°F and is stressed to 16 ksi. Performing the calculation shown above will result in a rupture life of 134,900 hr (15 years). The remaining life can also be determined by a method discussed in Section 2.28.

Assume that the bar in the furnace was operated at 16 ksi and 1100°F for 10 years. A problem occurs and for some reason the temperature rises to 1150°F for one year, with the stress remaining at 16 ksi. The question is: How much life remains? The life can be calculated for 1150°F as it was for 1100°F, and it turns out to be 3.61 years.

$$\sum \frac{\text{hours in operation}}{\text{hours to rupture}} = 1$$
$$\frac{10}{15.4} + \frac{1}{3.61} = 1$$
$$0.93 = 1$$

This shows that only $1 - 0.93$, or 7% of the life remains. Here 65% of the life was used by the 1100°F condition, and just raising the temperature by 50°F, to 1150°F, consumed 28% more of the life. This illustrates how sensitive creep life calculations are. It is very important that techniques such as those shown in API 530 be used for design, or large errors can result from poor data. One point should be obvious from the example: that small changes in temperature can result in large changes in the rupture life.

One precaution is that the stresses at temperature should be calculated and checked against the tensile and yield stresses as well as the creep life. Also, it should be mentioned that the methods in Ref. [2] are based on life in the range 20,000 to 200,000 hours; unreliable estimates can occur when shorter or longer life calculations occur.

It should be evident to the reader that whereas this example was concerned with axial stress in a bar, the creep could also have been due to pressure in a hot pipe such as a furnace tube. Here the concern would have been that the hoop stress would cause a creep rupture with time.

2.2.9   Shell Stresses

The stresses in cylindrical and spherical shells are of major importance in the petrochemical industries. Tanks holding millions of gallons of product, pressure vessels, towers, low-pressure spheres, thin pipes, and many other containers are classified as shells and can be analyzed as such. Figure 2.7 illustrates a tank shell that could be closed or open.

In shells, the thickness is considered small if it is less than 10% of the radius. The stresses are membrane-type stresses, which means that bending stresses are considered negligible. The bending stress shown on the figure is explained later in this section.

The S_1 and S_2 stresses are called by several names, which can be confusing. For example, S_1 is known as *longitudinal* or *meridian stress* and S_2 is known as *hoop*, *circumferential*, or *tangential stress*. In this section S_1 is called the longitudinal stress and S_2, the hoop stress.

In the following equations, p is the uniform internal or external pressure (psi), t is the shell thickness (in.), and R is the radius (in.). A plus sign for S_1 or S_2 indicates a

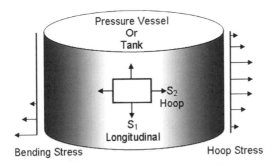

FIGURE 2.7   Tank and pressure vessel shell stresses.

tensile stress and can be thought of as an elongation of the element. Similarly, a negative sign is a compressive stress, as would occur with a vacuum or an external pressure.

Membrane stresses in a cylindrical vessel with closed ends with internal pressure are as follows:

Longitudinal: $S_1 = \dfrac{pR}{2t}$ lb/in^2

Hoop: $S_2 = \dfrac{pR}{t}$ lb/in^2

It should be mentioned that if a tank were open on top, S_1 would be essentially zero since there is no pressure to elongate the tank in that direction. For open tanks such as liquid storage tanks, an equation for the design hoop stress that is sometimes used is

$$S_2 = \frac{2.6 D_{ft}(H-1)(\text{S.G.})}{t \times \text{weld efficiency}} \qquad \text{lb/in}^2$$

H (ft) is the design liquid level in the tank, S.G. is the specific gravity of the liquid, D_{ft} is tank diameter (ft), t is the tank thickness (in.), and the weld efficiency is 1 if full inspection has been performed or 0.7 if the weld efficiency is unknown. Although design codes such as API 650 for new tank design and API 653 for in-service tanks should be followed, allowable values for S_2 are 0.8 times yield for the first course near the tank bottom and 0.88 times yield for the rest of the courses.

Membrane stresses in a sphere are

$$S_1 = S_2 = \frac{pR}{2t} \qquad \text{lb/in}^2$$

In all of the cases of membrane stresses above it should be remembered that they apply only well away from discontinuities. Such a condition might occur at a weld junction, such as at the bottom of a tank to the shell wall. Another junction would be the head of a vessel to the shell. A rough rule of thumb is to use only the equation for stresses $2.5(Rt)^{1/2}$ away from a discontinuity. For example, on a tank with a 400-in. radius and a $\frac{3}{8}$-in. shell wall thickness, the stress value S_1 or S_2 is valid approximately 31 in. away from the bottom. Figure 2.7 shows how the bending stresses start to come into effect as the calculated stresses approach the junction. The S_2 hoop stress becomes small and the bending stress becomes large, so that the stress of pR/t alone is no longer the controlling stress.

2.3 PIPING THERMAL FORCES, MOMENTS, AND FREQUENCIES

In the design of piping systems, a flexibility analysis of the piping is important to ensure that excessive thermally induced loads are not imposed on the connected machinery or structures. In piping systems there should be enough length and bends to make the system flexible enough so that loads are limited to acceptable values. The analysis of piping systems is a specialized area that utilizes computer programs to solve the maze of piping normally encountered in a process design. The programs analyze pipe hanger designs, spring supports, support settlement or growth, earthquake effects, misalignment, and a large number of other parameters. Most programs compare the loads with appropriate allowable loads and stresses from the various piping, pressure vessel, and machinery codes.

In analyzing piping, some engineers assume that piping smaller than 2 in. in diameter is flexible and do not include it in the analysis. Making such an assumption can greatly simplify complex systems. When the remaining system is larger piping, such assumptions seem appropriate. Calculating flexibilities by hand isn't done much anymore since computer solutions are readily available. What is important is to understand when a piping analysis is required.

The two piping arrangements shown in Figure 2.8 are a fixed L-bend subjected to a thermal expansion of 430°F temperature rise on 4-in. Schedule 40 pipe. The fixed wall could be other vessels or supports. Many situations can be approximated by this case, and it should also be useful to illustrate the basics of a piping flexibility analysis. Also shown in Figure 2.8 is a pipe with no bend. A pipe connecting two vessels or towers might be represented by this case.

In these examples, only the axial force in the x direction is shown; the moments are not shown. Notice how much flexibility the leg affords the system. It almost eliminates the axial force. Of course, it does this by introducing sizable moments at the supports that must also be considered but are not shown.

When reviewing the calculated loads and moments and to determine if they are acceptable, allowable values can be obtained from:

- The equipment manufacturer
- For compressors, API 617

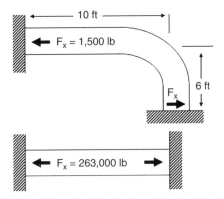

FIGURE 2.8 Thermal piping loads with and without a bend.

TABLE 2.2   Piping Frequency Constants

$f_n = \alpha\left(\dfrac{EI}{WL^3}\right)^{1/2}$	Cycles/sec
Cantilever	$\alpha = 0.265$
Simply supported (SS)	$\alpha = 0.743$
Fixed ends	$\alpha = 1.66$
Fixed end and SS	$\alpha = 1.16$

- For pumps, API 610
- For pressure vessels, ASME pressure vessel codes

The following approximation for allowable loads and moments on industrial machinery is shown because of its simplicity; it is used only to see if a value seems reasonable, not necessarily that it be acceptable. It does not always result in conservative values and should be used cautiously, and certainly not on air-finned coolers or other thin vessels or structures.

- Any resulting moments should be less than 200D ft-lb.
- Any resulting forces should be less than 100D lb.

where D is the nominal pipe size (in.).

Four other calculations are useful for determining the natural frequency of piping spans. The basic equation for the first vibration mode is shown in Table 2.2. In this equation E is in lb/in^2, I in in^4, W is total pipe weight (lb), and L is span length between supports (ft). Table 2.2 are constants for various end conditions.

The frequencies calculated can be compared with possible excitation sources such as rotating machinery unbalance and pressure pulsations. The magnifier methods described in Chapter 3 can be used to calculate amplitudes if resonance frequencies cannot be avoided.

2.3.1   Piping Failures

Piping systems are usually quite complex in the petrochemical and power industries, and troubleshooting failures and implementing solutions can occupy a significant portion of an engineering department's time. Typically, failures can be related to corrosion or to erosion of the piping, which is usually directed to the materials group. Pipe thicknesses required to contain pressure can be determined by considering the equations used in the ASME code. These equations are for pressure vessel shells but are adequate for the type of noncode piping checks with which we are concerned.

$$t = \frac{pR}{Se - 0.6p} + \text{corrosion allowance}$$

TABLE 2.3   Weld Efficiencies

Butt joints welded the same amount each side:

Full radiography	$e = 1.00$
Spot examined	$e = 0.85$
Not examined	$e = 0.70$

Single-welded butt joint with backup strip:

Full radiography	$e = 0.90$
Spot examined	$e = 0.80$
Not examined	$e = 0.60$
Most other lap joints	$e = 0.45–0.6$

where

$$t = \text{minimum required shell thickness (in.)}$$

$$p = \text{internal pressure (lb/in}^2\text{)}$$

$$R = \text{vessel inside radius (in.)}$$

$$S = \text{allowable stress (17,500 lb/in}^2 \text{ is typical)}$$

$$e = \text{joint efficiency}$$

Typical joint efficiencies are defined in Table 2.3.

Here's an example to determine the shell thickness for a horizontal pressure vessel that will have an operating pressure p of 100 psig, a radius of 48 in., a corrosion allowance of 0.125 in., and that is made from SA-515-70 plate with $S = 17,500$ lb/in^2. The joint is butt welded and will be spot examined.

$$t = \frac{pR}{Se - 0.6p} + \text{corrosion allowance}$$

$$= \frac{100(48)}{17,500(0.85) - 0.6(100)} + 0.125$$

$$= 0.325 + 0.125 = 0.45 \text{ in.} \quad \text{use } \tfrac{1}{2}\text{-in. plate}$$

Notice the versatility of this equation. By increasing the inspection technique to 100% radiography on this joint, one can calculate how much additional pressure the shell is designed for. Such a condition could occur in a re-rate case if no corrosion has occurred.

Many older vessels are inspected with modern techniques but may have been designed to an older code. Many inclusions, cracks, and other defects will be found and have to be ground out and repaired or analyzed by fitness-for-service methods to determine if they are fit for service. These vessels may have been designed with an efficiency $e = 0.6$, which means that they were acceptable per code at the time of manufacture without examination. In some cases welding rods were found in the root passes and were used as filler material. This is not pretty but is a fact of the time. Unfortunately, once inspected and found, even though the device has been in successful service for many years, a decision has to be made and risk assessments performed. On some large liquefied petroleum gas (LPG) spheres this necessitates grinding out and welding up hundreds of defects.

Failures related to fatigue usually appears as cracks in highly stressed areas. Typically, these occur in areas of poor welds where the stress concentration can be as high as 4, compared to a proper weld, which usually has a value below 2. Unsupported small piping is especially prone to cracking. Bracing such pipes helps.

Fatigue failures result from cycling stresses, usually due to piping vibration. These types of failures will appear similar to those discussed in Section 2.25. Low-frequency piping vibrations up to about 100 cycles/sec should be corrected when they are above approximately 20 mils peak to peak (p-p). Higher frequencies should be controlled when they reach 10 mils p-p. Localized stiffening by bracing is a typical fix used. Temporary braces can be wedged in to see if they help.

Low-frequency vibration of piping can be caused by piping resonance due to inadequate piping support or machinery such as pumps or reciprocating compressors. Higher-frequency vibration of piping can be caused by fluid dynamics in the piping, which may need to be tuned out by using pulsation bottles. It can also be due to piping or structure resonance.

Leakage at flanges results in leaks, which are considered a piping failure. Leakage can be due to loose bolts caused by vibration. Stopping the vibration should solve the problem. Leaking flanges can also be due to excessive piping forces which overload the flange bolting and cause gasket leaks. A piping flexibility analysis should identify such problems. Proper piping designs to ensure flexibility should eliminate this problem.

Heavy metallic knocking of the piping may be caused by surging or water hammer of the fluid within the piping. It can be caused by rapid opening and closing of improperly tuned control valves, or pump startup and shutdown. Forces can be high enough to knock piping off the racks and break support clamps. In Chapter 4, water hammer and pressure surges and the forces produced are discussed in a simplified manner.

Since many leaks and fires occur at flanges, the guide in Table 2.4, has been used on flanges that have been defined as *critical flanges*, usually flanges that have been prone to leakage and are left uninsulated if safety and the process allow. When safety is a problem, guard cages are sometimes used so that the flanges can be inspected easily and the bolts checked for tightness.

TABLE 2.4   Critical Flange Guide

Step	Item
A	Check studs and nuts. Run nuts onto studs by hand to check fit.
B	Lubricate studs and nuts, including the nut face, with suitable thread lubricant (e.g., Never-Seez).
C	Clean and check flange gasket contact faces.
D	Clean and check flange nut bearing surfaces.
E	Check flange-to-flange face alignment $\frac{1}{32}$-in. maximum gap.
F	Check gasket insertion gap 0 to $\frac{1}{8}$-in. maximum gap.
G	Check flange-to-flange parallel alignment $\pm \frac{1}{64}$-in. maximum.
H	Install lower half studs.
I	Inspect gasket.
J	Insert gasket.
K	Install remaining studs and nuts and hand-tighten. Ensure that there is full thread engagement at each nut.
L	Mark bolting sequence numbers.
M	Tighten bolts with hand wrenches at 5 to 20 ft-lb using a crisscross pattern. Qualifying mechanics on the correct procedure for tightening critical joints is a good idea.
N	Tighten bolts up to 40% of final torque values in a crisscross pattern.
O	Tighten bolts up to 70% of final torque values in a crisscross pattern.
P	Tighten bolts up to final torque values in a crisscross pattern.
Q	Use a rotational pattern (e.g., $1, 2, 3, \ldots$) starting with stud 1 and continue clockwise around the flange. Continue until no more nut movement is observed.

2.4   ALLOWABLE AND DESIGN STRESSES

Allowable stresses are determined by recognizing which of the various possible failure modes apply to the problem being analyzed. The strength of the material is then utilized from tabulated test data. Primary failure modes considered in engineering are discussed next.

Ultimate Strength S_u   Here the part will crack and fracture. In a tensile machine, the part breaks. For steel, a rough approximation is 500 times the Brinell hardness number (BHN).

Yield Strength S_y   In this mode the material will not return to its original shape. In a tensile machine, a ductile part would "neck down." If interference or fit-up were a problem, this would be an important failure mode. In general, it's a good idea not to design near the yield strength.

Shear Strength S_s Shafts fail in shear when there is an excessive torque overload. Shear strength data are more difficult to find. If data are unavailable, use the maximum shear theory:

$$S_s = \tfrac{1}{2} S_y$$

Endurance Strength S_e Of the failures the author has analyzed, 90% have been fatigue failures due to some form of cyclic loading. In the absence of data for steel, use

$$S_e = \tfrac{1}{2} S_u$$

2.5   FATIGUE DUE TO CYCLIC LOADING

The endurance limit for steel and many metals is important. If the part can cycle through 10 million stress cycles at a given stress level, it can be said with some authority that the part will not fail in fatigue no matter how many more cycles it undergoes.

Figure 2.9 shows the endurance limit for steel and how it is determined. The material is placed in a test machine and undergoes cyclic stress at a given level and run to failure. Note that after 10^7 cycles there is an upper stress at which the material can be cycled and no failure occurs. This is called the *endurance limit* and can be used to determine if a part will fail in fatigue. In reference to Figure 2.9, if we put $\pm 70{,}000$ lb/in^2 bending stress on a shaft, we would expect it to fail. However, a failure would not be expected at less than 50,000 lb/in^2 no matter how many cycles it undergoes. Some materials do not have an endurance limit as just defined and might behave more as the straight line shown.

When only static stresses are to be considered, the acceptability of the calculated stresses could be determined easily. All one would have to do is to see if a stress value exceeded one of the failure criteria. But what happens when a fluctuating stress is superimposed on a static stress? The endurance limit alone is of little use. For this

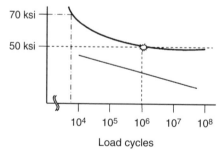

FIGURE 2.9   Endurance limit for steel.

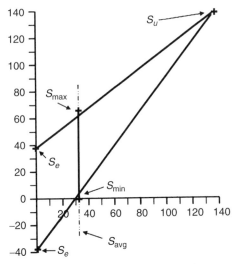

FIGURE 2.10 Modified Goodman diagram.

type of failure a diagram called the *modified Goodman* is useful and is shown in Figure 2.10.

The steps for construction are simple once the stresses have been calculated. The values shown are for a case history used later in this chapter.

1. S_e = endurance strength with adjustments for corrosion, size, and so on = 37,000 lb/in².
2. S_u = ultimate strength = 140,000 lb/in².
3. $S_{avg} = \frac{1}{2}(S_{max} + S_{min}) = \frac{1}{2}(64,000 + 0) = 32,000$ lb/in².
4. S_{max} = maximum fluctuating stress = 64,000 lb/in².
5. S_{min} = minimum fluctuating stress = 0 lb/in².

When data are not available and a quick check on the fatigue data is needed, the following can be used for steel at 10^6 cycles. S'_e is the actual polished rotating bending specimen endurance limit. When that is not known, an estimate for steel is one-half the ultimate tensile strength. The surface finish factor C_S data shown on Table 2.5 is for steels with tensile strengths in the range 100,000 to 200,000 lb/in²:

$$S_e = C_L C_D C_S S'_e$$

Normal preference is to put the stress concentration factors on S_{max} and S_{min}. However, with highly corroded surfaces, one can neglect any stress concentration factors in a first pass.

TABLE 2.5 Endurance Limit Factors

		Factor	
Condition	C_L	C_D	C_S
Bending	1.00	0.9 < 2 in. diameter	
		0.7 > 4 in. diameter	
Torsional	0.58	0.9 < 2 in. diameter	
		0.7 > 4 in. diameter	
Axial	0.90	1.0	
Mirror polish			1.0
Fine grind			0.9
Machined			0.7
Hot rolled			0.5
As forged			0.3
Corroded			0.2

It is important to realize that almost 90% of the fatigue life goes into producing a crack. Once the crack is established, not much remaining life is left. That is why the various surface finishes so greatly affect the endurance limit. Surface defects are produced which reduce the fatigue life. Figure 2.11, a graph of the fatigue life remaining when a crack is present, illustrates this point. Fretting corrosion, which is a form of wear, results in small cracks which greatly reduce the fatigue strength and may cause shaft failures (discussed in more detail in Section 9.14). The graph illustrates a cyclic axial stress on a crack in a steel bar 1 in. thick. Life is considered to be complete when the initial crack grows through the bar. The graph illustrates that an initial crack of any size can drastically reduce the fatigue life of a part. The larger the initial crack, the shorter the life.

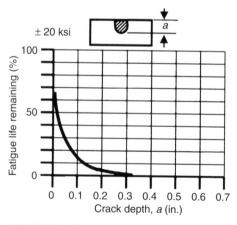

FIGURE 2.11 Remaining life with a crack.

2.6 ELONGATIONS AND DEFLECTION CALCULATIONS

The elongation or stretch of a structure due to the applied load P (lb) is shown in Figure 2.12. In equation form this elongation is given by

$$\delta_P = \frac{PL}{AE} \quad \text{in.}$$

where $E_{steel} = 30 \times 10^6$ lb/in^2 and $A_{cross\ section}$ is in in^2. Strain is defined as

$$\varepsilon = \frac{\delta_P}{L} = \frac{P}{AE} = \frac{S_A}{E} \quad \text{in./in.}$$

Two frequently used deflection-of-beam equations for a simply supported beam and a cantilever beam are shown in Figure 2.13. Tables of these types of diagrams, together with the moments, are readily available in reference books [4]. The case of a simply supported beam with built-in or fixed ends is also shown since in some

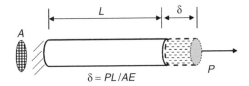

$$\delta = PL/AE$$

FIGURE 2.12   Elongation due to axial load on a bar.

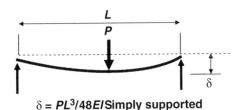

$\delta = PL^3/48EI$ **Simply supported**

$\delta = PL^3/192EI$ **Fixed**

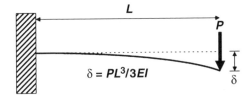

$$\delta = PL^3/3EI$$

FIGURE 2.13   Simply supported and cantilever beam deflections.

cases, such as bolted connections or bearings, the end supports are somewhere in between. For the fixed case the moment at the wall and at the load is

$$M_{wall} = M_{load} = \frac{PL}{8}$$

For the free, simply supported case,

$$M_{load} = \frac{PL}{4}$$

Slope at simply supported beam:

$$\theta_{support} = \frac{PL^2}{16EI} \quad \text{rad}$$

Slope in terms of center deflection simply supported beam:

$$\theta_{support} = \frac{3\delta}{L} \quad \text{rad}$$

For the cantilever case,

$$M_{wall} = PL$$

A general equation with three loads applied that is useful in bearing problems is included in Figure 2.14.

One other beam equation that is quite useful is the guided cantilever (Figure 2.15). In this equation the slope at the tip is forced to remain zero. It can represent a beam

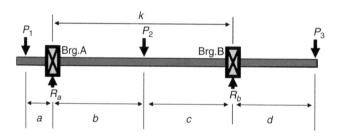

$$R_a = [P_1(k + a) + (P_2c) - (P_3d)] / k$$
$$R_b = [(-P_1a) + (P_2b) + P_3(k + d)] / k$$

FIGURE 2.14   General reaction equation for a shaft.

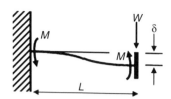

FIGURE 2.15 Guided cantilever beam equation.

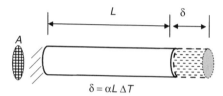

$$\delta = \alpha L \, \Delta T$$

FIGURE 2.16 Elongation of a bar due to thermal growth.

forced to move vertically in a bushing so that it cannot rotate. The moments can then be used for stress calculations.

$$M = \frac{WL}{2}$$

$$\delta = \frac{WL^3}{12EI}$$

When a deflection is forced, the moment due to it is

$$M = 6EI \frac{\delta}{L^2}$$

Units can be compatible units: $I = \text{in}^4$ and $E = \text{lb/in}^2$.

Due to thermal growth (the temperature of structure differs from ambient by ΔT) (Figure 2.16), the expansion or elongation of the bar is

$$\delta_t = \alpha L \, \Delta T$$

$$\alpha_{\text{steel}} = 6.6 \times 10^{-6} \, \text{in./in.-°F}$$

$$\alpha_{\text{aluminum}} = 13 \times 10^{-6} \, \text{in./in.-°F}$$

To appreciate the versatility of these two equations, consider the following example. Bearings on a motor have been failing, and on disassembly both bearings are found to be tight on the shaft and in the housing. The question is: Could a high load be produced by thermal growth of the shaft?

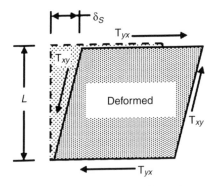

FIGURE 2.17   Shear deformation.

Using the two equations above, we can say that thermal growth δ_t is restrained fully by load P, or P compresses the rod by δ_l:

$$\delta_t = \delta_L$$

Substituting the equations results in

$$P = \alpha A E\, \Delta T$$

For a 3-in.-diameter steel shaft and ΔT of 50°F,

$$P = \frac{6.6 \times 10^{-6} \pi\, (3^2)(30 \times 10^6)(50)}{4}$$

$$= 70{,}000\ \text{lb}$$

This amount of load can overload a bearing that is not designed for heavy thrust loads. The bearing has to be free in the housing or excessive loads will develop.

One other deformation is shearing strain, which is due to shear loads and is shown as in Figure 2.17. Shear strain is defined as

$$\gamma = \frac{\delta_s}{L}$$

2.7   FACTOR OF SAFETY

In mechanical engineering a definition frequently used for the *factor of safety* (FS) is that it is the load or stress a structure is capable of withstanding while still performing its function, divided by the *applied load* or stress. The applied load is sometimes called the *working load*. The term *design margin* is also used and has the same meaning but is preferred since it eliminates the word *safety*, which implies that all is well. For example, for a uniaxial sample in tension,

$$FS = \frac{S_y}{S_{applied}}$$

With this ratio equal to 2, the applied stress could be doubled before the part would yield.

Materials with elongations over 5% are called *ductile*. In ductile materials the factor of safety is based on the yield strength, as in the equation above, or the endurance limit. For elongations of less then 5%, material is usually considered *brittle*, and because such material fractures without yielding, the ultimate strength is used instead of the yield strength.

With static or alternating loads only, calculation of the factor of safety is fairly straightforward. When yield stress is used, the FS value used is generally between 1.5 and 4. With well-defined design loads, reliable material properties, and stable operating conditions, lower values usually apply, with 2 typically being used. When dealing with brittle materials, many designers double the factors of safety they would use on ductile materials or apply fracture mechanics techniques. When dealing with materials with low toughness, in critical applications where brittle fracture is of concern, the foregoing FS approach can produce erroneous and nonconservative results and should not be used.

In fatigue analysis with alternating stress cycling about an average (mean) stress, using the geometry on a Goodman diagram allows the FS to be calculated. This involves determining when the stress range falls outside the Goodman diagram envelope. Thus, with an FS value below 1, fatigue failure in a ductile material could be expected, in fewer than 1 million cycles.

For combined loading such as bending (S) and torsion (τ) in a ductile material, the following FS in fatigue is used [5].

$$FS = \left[\left(\frac{kS_{alt}}{S_e} + \frac{S_{avg}}{S_y} \right)^2 + 3\left(\frac{k\tau_{alt}}{S_e} + \frac{\tau_{avg}}{S_y} \right)^2 \right]^{-1/2}$$

Notice that k is the stress concentration factor and is applied to the alternating stress component only. For simplification, the same k is shown for the bending and torsional components. With axial or bending stresses only, the equation simplifies to

$$FS = \frac{1}{kS_{alt}/S_e + S_{avg}/S_y}$$

As an example, consider a small shaft that has an average and alternating bending load, along with an average torsional load, with no torsional alternating load. Such might be the case of a pulley with a V-belt on the end of a motor shaft. For each shaft rotation, a point on the shaft sees a bending load and some average load due to belt tension, along with some average torque load. The desired result is the FS of the shaft in a keyway region, which has a stress concentration k of 1.65. The data needed for the calculations are

$$S_{alt} = 6.15\,ksi \quad \tau_{alt} = 0 \quad\quad S_e = 53\,ksi$$
$$S_{avg} = 1.50\,ksi \quad \tau_{avg} = 1.5\,ksi \quad S_y = 110\,ksi$$

After performing the calculations, the FS value is 4.9.

2.8   CASE HISTORY: AGITATOR STEADY BEARING LOADING

Agitators are devices that mix and blend various products. Designing for the degree of agitation is a fairly complex procedure; however, most of the better known vendors have a wealth of experience to help potential customers. In this case we are only interested in the loading on the vessel. This is because in many fast-moving projects, a vessel design may lead the agitator selection by months. By providing maximum agitator loads that the vessel can expect, correct reinforcement can be added early in the design. Late design changes are quite prohibitive from a project manager's view.

The purpose of this case is to show the technique used to calculate the steady bearing load as it uses load and deflection equations that we have just reviewed. A similar procedure is used to calculate the vessel nozzle loads R_2, M_2, and the torque reaction (Figure 2.18). In addition, the thrust reaction and dead weight are considered. Only the lateral steady bearing load R_1 is being reviewed. The vessel designer will convert this into a bending moment and calculate the head thickness and reinforcement required.

The steady bearing has a radial clearance of 0.020 in., so two calculations are required. First, the problem is that of a cantilever beam until the shaft moves through

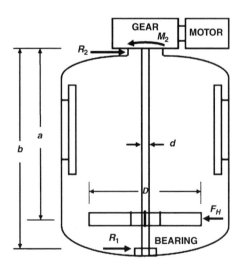

FIGURE 2.18   Agitator reaction loads.

the clearance. When the shaft makes contact, the shaft becomes simply supported at the steady bearing. R_1 is therefore the required load on the steady bearing. The applied load supplied by one agitator manufacturer is

$$F_H = \frac{19,000f(\text{hp})}{ND} \qquad \text{lb}$$

where

$$f = \begin{cases} 1 & \text{ideal operating conditions} \\ 4 & \text{poor operating conditions} \\ 5 & \text{possible impacting conditions} \end{cases}$$

$$\text{hp} = \text{horsepower}$$
$$N = \text{rpm}$$
$$D = \text{diameter (in.)}$$

Then

$$F_H = \frac{19,000(5)(30)}{100(48)} = 594 \text{ lb}$$

To deflect through the bearing clearance, the cantilever beam equation is used:

$$P = \frac{3EIy_{max}}{L^3} = \frac{3(30 \times 10^6)(20.1)(0.02)}{(176)^3}$$
$$= 7 \text{ lb}$$

Obviously, the shaft is very flexible, since it takes only 7 lb to contact the bearing. Only the second part, simply supported at the steady bearing, need be considered.

The load and reaction case shown on the agitator drawing can be found using the general reaction equation shown in Figure 2.14. Since some loads are zero and $P_2 = F_H$, we have

$$R_b = \frac{P_2 b}{k}$$

Using the nomenclature for this problem gives us

$$R_1 = \frac{F_H a}{b} = \frac{594(158)}{176}$$
$$= 533 \text{ lb}$$

This is what will be given to the vessel designer as a worst-case steady bearing load.

2.8.1 Additional Agitator Guidelines (Single Impeller)

Equations that are useful for operating or analyzing agitator failures are as follows [6]:

Baffle loading:

$$F_B = \text{force on baffle}$$

$$= \frac{21{,}000 \text{ hp}}{\text{no. baffles} \times \text{tank diameter (ft)} \times \text{rpm}}$$

Horsepower requirements:
    Pitch blades:

$$\text{hp} = \left(\frac{D}{394}\right)^5 (\text{specific gravity}) \, (\text{rpm})^3$$

Straight blades:

$$\text{hp} = \left(\frac{D}{320}\right)^5 (\text{specific gravity}) \, (\text{rpm})^3$$

Typical design ratios:

$$\frac{\text{tank diameter (ft)}}{\text{impeller diameter (ft)}} \qquad 3$$

$$\frac{\text{tank diameter (ft)}}{\text{impeller distance bottom tank (ft)}} \qquad 3$$

$$\frac{\text{tank diameter (ft)}}{\text{baffle width (ft)}} \qquad 12$$

$$\frac{\text{tank diameter (ft)}}{\text{liquid depth minimum (ft)}} \qquad 1$$

Supporting structure design: The structure should be rigid enough so that the agitator support plane doesn't tilt more than 0.15° in any direction. Some maximum reactions on the support plane for design use are

$$\text{moment } M = 0.25 F_H \, [\text{shaft length (ft)}] \qquad \text{ft-lb}$$

$$\text{torque } T = 13{,}125 \text{ hp/rpm} \qquad \text{ft-lb}$$

Shaft stresses on agitators:

Shear stress due to torque for 304 and 316 stainless steels:

$$\tau = 6000 \text{ lb/in}^2$$

Axial stress due to shaft and impeller weight:

$$\sigma = 10,000 \text{ lb/in}^2$$

This accounts for dynamic loads, keyways, and so on. Allowables may be extended to other materials by the ratio of the yield strengths.

2.9 CASE HISTORY: EXTRUDER SHAFT FAILURE

In this example a shaft had failed in a piece of extrusion equipment. The failure mode was that the shaft had twisted nearly 30° at its input end. This is the end of the shaft that saw the full torque. Over the years the motor size was increased from 800 hp to 1200 hp. The shaft material was 316 stainless steel. The plant personnel had success with 17-4 PH 1025 (Table 2.6) and the question was whether a change in material would help and why the shaft had twisted.

$$\text{Shaft diameter } (D) = 5.125 \text{ in.}$$

$$\text{Keyway fatigue stress concentration factor } (K) = 2.55$$

$$\text{Shaft speed (rpm)} = 200$$

1. Average shear stress:

$$\tau_{avg} = \frac{321,000(\text{hp}) \text{ (S.F.)}}{D^3(\text{rpm})} \quad \text{lb/in}^2$$

2. Alternating shear stress (τ_{alt}) is 5% of the mean based on motor power fluctuations.
3. Safety factor in fatigue:

$$FS_f = \frac{1}{1.73(K\tau_{alt}/S_e + \tau_{avg}/S_y)}$$

where $\tau_{alt} + \tau_{avg} < 0.5S_y$.

TABLE 2.6 Material Properties at 600°F (ksi)

	S_y	S_e
17-4 PH 1025	132	85
316 (typical)	40	37

TABLE 2.7   Extruder Shaft Failure

Horsepower	Service Factor	Material	FS_f	FS_y
800	1.15	316	1.85	1.74
1200	1.25	316	1.13	1.07
1200	1.25	17-4 PH	3.56	3.51

4. Calculate the safety factor in yield:

$$FS_y = \frac{0.5S_y}{\tau_{alt} + \tau_{avg}}$$

Table 2.7 summarizes the results.

It is evident that the original shaft design with the 800-hp motor was adequate. It had operated for 10 years with no twisting of the shaft or cracking in the keyway. This is not the case with the up-rate, as the safety factors are marginal. Because of its low yield strength, any overload could twist the shaft, as indeed it did. A shaft of 17-4 PH 1025 material solved the problem.

2.10   DYNAMIC LOADING

2.10.1   Centrifugal Force

Newton's second law is used extensively in engineering. It states that a change in motion (acceleration) is proportional to the force applied:

$$F_a = ma$$

Since $m = W/g$,

$$F_a = \frac{Wa}{g}$$

This is the force required to accelerate W due to a. Also useful is the nomenclature

$$G's = \frac{a}{g}$$

so

$$F_a = W \times G's$$

Two G's is not an unusual impact type of acceleration, such as that which occurs when one railroad train is coupled to another.

This equation can be developed into force due to centrifugal effects. This is the radial force produced by a rotating mass, a weigh on a string, or swinging in a circle. When a weight is swung on a long rope, this is the force in the hand holding the rope. In this case, where ω is in rad/sec,

$$a = R\omega^2$$

$$\omega = \frac{2\pi}{60} \, (\text{rpm})$$

$$F_c = \frac{W}{g} \times R \left[\frac{2\pi}{60} (\text{rpm}) \right]^2$$

Where the acceleration is $a = R \, [(2\pi/60)(\text{rpm})]^2$,

$$F_c = 28.4 \times 10^{-6} \, WR(\text{rpm}^2) \qquad \text{lb}$$

$$= 28.4WR \left(\frac{\text{rpm}}{1000} \right)^2$$

where W is in pounds and R is in inches. R can also be eccentricity ε (in.), like a radial runout:

$$F_c = 28.4W\varepsilon \left(\frac{\text{rpm}}{1000} \right)^2$$

Let's look at an example. Figure 2.19 shows a 3000-lb rotor rotating at 300 rpm with a radial runout of 0.010 in. Was this excessive? One rule of thumb is that a reaction force that is 10% of the rotor weight is about equivalent to a grade 6.3 rotor imbalance, which is pretty good.

$$F_c = 28.4W\varepsilon \left(\frac{\text{rpm}}{1000} \right)^2$$

$$= 28.4(3000)(0.010) \left(\frac{300}{1000} \right)^2 = 77 \text{ lb}$$

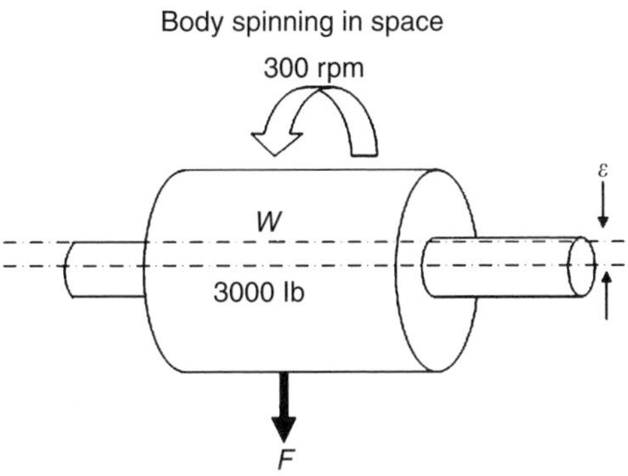

FIGURE 2.19   Rotor with runout.

Since 10% of the 3000-lb rotor is 300 lb, the 77-lb resulting imbalance appears acceptable.

2.10.2   Inertias and WR^2

The torque required to accelerate or decelerate can be determined similar to force:

$$T = J\alpha$$

where T (ft-lb) is the torque to accelerate, J (ft-lb/sec²) is the mass moment inertia, and α is the angular acceleration (rad/sec²). J is tabulated in mechanics handbooks for various shapes.

One very useful equation is for a cylinder about its center:

Solid cylinder:

$$J = \frac{1}{2}\frac{W}{g}r^2$$

Hollow cylinder:

$$J = \frac{1}{2}\frac{W}{g}(r^2_{outside} - r^2_{inside})$$

Useful angular accelerations with ω in rad/sec and t in seconds:

$$\alpha = \frac{\omega_{final} - \omega_{initial}}{t}$$

In terms of rpm,

$$\alpha = \frac{rpm_{final} - rpm_{initial}}{t}\frac{2\pi}{60} \qquad rad/sec^2$$

A term that appears frequently in work with rotating machinery (e.g., motors, compressors, drivetrains) is WR^2. To relate this term to what has just been discussed, if the J term in $T = J\alpha$ is to be identical, divide WR^2 by the acceleration of gravity g (ft/sec²), where W is the weight in pounds and R is the radius of gyration.

For a solid cylinder about its own axis, where r is the cylinder radius,

$$R^2 = \frac{1}{2}r^2 \qquad ft^2$$

The torque equation can be rearranged into the useful form

$$T = \frac{WR^2 \, \Delta rpm}{307.6t} \qquad lb\text{-}ft$$

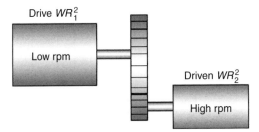

FIGURE 2.20   Effect of gear ratio on WR^2.

where T is the torque (lb-ft), Δrpm the change in rpm, t the time for change (sec), and WR^2 the resisting inertia (lb-ft²). Machinery manufacturers usually provide inertia information in terms of WR^2 (lb-ft²)—thus its importance. One thing about inertias (WR^2) is that the apparent inertia is affected by speed increasers or reducers by the square of the ratios (Figure 2.20).

The WR^2 is usually related back to the drive shaft:

$$WR^2_{1,\text{drive}} = \left(\frac{\text{rpm}_{\text{driven}}}{\text{rpm}_{\text{drive}}}\right)^2 WR^2_{2,\text{driven}}$$

Notice that if the driven unit is rotating faster than the input, this is a speed increaser and the driven WR^2 will appear larger to the drive side (i.e., the load to accelerate is higher at the drive). Similarly, if it is a speed reducer, it will appear lower to the drive side.

2.10.3   Energy Relationships

Energy relationships are valuable tools for analyzing engineering problems, especially for impact or shock loads. The important key is the use of conservation of energy:

$$\text{potential energy} = \text{kinetic energy}$$

$$\text{PE} = \text{KE}$$

For linear cases:

$$\text{KE} = \frac{1}{2}mV^2 = \frac{W}{2g}V^2 \qquad \text{ft-lb}$$

$$\text{PE} = Wh = \frac{1}{2}(K\delta^2) \text{ spring}$$

where W is in pounds, $g = 32.2$ ft/sec², V is in ft/sec, and h is in feet.

For angular cases:

$$KE = \frac{1}{2}J\omega^2$$

$$PE = T\theta$$

where T is in ft-lb, θ is in radians, ω is in rad/sec, and J is in ft-lb-sec^2.

Power relationships:
  Rotary horsepower

$$hp = T\frac{rpm}{63,000}$$

  Linear horsepower

$$hp = \frac{FV}{550}$$

where F is in pounds, V is in ft/sec, and T is in in.-lb.

2.11   CASE HISTORY: CENTRIFUGE BEARING FAILURES

This case history uses loading calculations and dynamic force equations. Figure 2.21 shows a free-body diagram (FBD) of a centrifuge. In operation, the feed comes in the top and is designed so that heavier particles are centrifuged out through the nozzles. The unit can be "washed" if plugging occurs.

In this case failures occurred on the bearings, and a third bearing was added at point C to take the cantilever effect of the bowl. Periodically, the unit would shake violently and would be washed. Lubrication, assembly errors, defective bearings, and other factors were all suspected. Due to washing and load imbalance, it was decided to look at the effect of imbalance on bearing life. The first step was to simplify the problem.

Since the imbalance will affect the left-to-right/right-to-left (x-axis) forces, only these need be considered. The important forces and reactions are shown on the free-body diagram:

$$\Sigma F = 0 \leftarrow +$$

After simplifying, we have

$$R_C = R_B - P \pm F_U$$

$$\Sigma M_{RB} = 0 \qquad + \text{ is cw}$$

$$-PL_B + R_C L_S \pm F_U(L_S + L_C) = 0$$

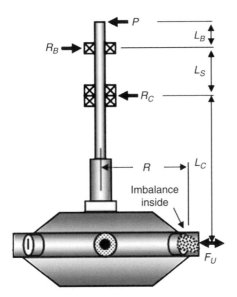

FIGURE 2.21   Centrifuge imbalance.

After simplifying and substituting R_C, we obtain

$$R_B = \frac{P(L_B + L_S) \pm F_U(2L_S + L_C)}{L_S}$$

Substituting R_B into the sum of the forces equation and solving for R_C yields

$$R_C = \frac{PL_B}{L_S} \pm F_U\left(3 + \frac{L_C}{L_S}\right)$$

Since the other parameters are known, all that is left to do is to define P and F_U. P is the belt load [7] and for our case is simplified to D_{pulley} (in.)

$$P = \frac{303{,}000(hp)}{D_{pulley}(rpm_{pulley})} \qquad lb$$

This is for a belt tensioned to the manufacturer's limits. Overtensioning by the mechanic can increase this by a factor of 3 or more.

The unbalanced product force (F_U) is due to an unbalanced weight (product plug) W at a radius R:

$$F_U = \pm 28.4 WR\left(\frac{rpm}{1000}\right)^2 \qquad lb$$

TABLE 2.8   Bearing Life

Condition	R_B (lb)	B_{10} life[a]
1. No unbalance, due to product, only assembly balance $W = 0.01$ lb at R	1800	Very long
2. $W = 0.1$ lb at R	3800	2.3 years
3. $W = 0.2$ lb at R	7200	3 months

[a]For a discussion of the B_{10} life, see Section 2.18.

"Plus or minus" is used since the one that causes the largest reaction will be used. After going through the mathematics, we have

$$P = 1000 \text{ lb}$$

$$F_U = 3200W \quad \text{lb}$$

The R_B bearing is a SKF N 315 roller bearings. Assuming imbalances and calculating bearing life from the manufacturer's product directory, the data in Table 2.8 are obtained.

From the analysis it was determined that the assembly tolerances are critical and that no product unbalance is allowable and a wash schedule must be adhered to. The analysis allowed limits to be set on tolerances to keep the load within range. A larger bearing at the R_B location was considered, but space was a factor and product imbalance could be large enough to fail much larger capacity bearings.

2.12   CASE HISTORY: BIRD IMPACT FORCE ON A WINDSCREEN

This analysis illustrates use of the energy equations to solve an impact problem. The resulting equation is useful for many applications. In this example, a new, thicker windscreen (Figure 2.22) was to be installed on an aircraft. The purpose of the analysis was to determine the force produced on the polymer windscreen when a 1-lb bird struck it at 150 mph. This information could then be used in a static test to determine if the windscreen would fail. A value that is easily determined from tests is the spring constant K (lb/in.) of the windscreen. A load is put on the center of the windscreen and the deflection under the load is measured. This was determined to be 2000 lb/in.

The basis of this analysis is that the energy of impact is totally absorbed by the elastic distortion of the windscreen, and the model sketch is shown in Figure 2.23.

$$KE = PE$$

$$\frac{\frac{1}{2}WV^2}{g} = \frac{1}{2}K\delta^2$$

FIGURE 2.22   Aircraft windscreen.

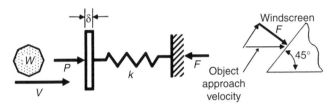

FIGURE 2.23   Impact of a bird.

Dividing both sides by δ, letting $K = P/\delta$, and simplifying gives us

$$\frac{WV^2}{g} = K\delta^2$$

But the force in the spring is $F = K\delta$, so $\delta = F/K$. Substituting and solving for F yields

$$F = V\left(\frac{KW}{g}\right)^{1/2}$$

This equation can be used to determine the reaction force F (lb) of a weight W (lb) moving against a spring member K (lb/in.) at a given velocity V (in./sec).

For this case, the windscreen is at an angle of 45°, so the components perpendicular to the windscreen need to be calculated and utilized in the equation.

$$V = 0.707(150) = 106\,\text{mph}$$

$$W = 0.707(1.0) = 0.7\,\text{lb}$$

$$F = 1866\,\text{in./sec} \; (2000\,\text{lb/in.} \times 0.7\,\text{lb/386 in./sec}^2)^{1/2}$$

$$= 3553\,\text{lb}$$

Obviously, this is a very conservative number since the bird does not expend all the kinetic energy when it hits the windscreen. Its terminal velocity is not zero when it moves off the windscreen at the 45° angle. Also, the spring constant of the bird is not taken into effect, which should be combined with the windscreen as an equivalent spring constant. Some validity of the results come from air cannon tests of frozen chickens conducted to evaluate the effect of bird strikes on commercial airline windscreens—a rather gory testing assignment to read, but useful. It shows that a 4-lb frozen chicken hitting a solid windscreen at 300 mph results in a 38,000-lb force. Our calculation would show a 17,000-lb force. But then again, their bird was frozen and their windscreen thick.

The equation will be useful in examining the relative effect of factors such as velocity, angle, weight, and spring constant on the reaction force, as well as verifying the order of magnitude for use in a finite-element analysis.

2.13   CASE HISTORY: TORSIONAL IMPACT ON A PROPELLER

The linear impact equation was shown in Section 2.12:

$$F = V \left(\frac{KW}{g} \right)^{1/2}$$

In a similar manner the torsional impact equation can be developed:

$$T = \frac{\pi(\text{rpm})}{30} (JK)^{1/2}$$

$$J = \frac{WR^2}{g}$$

In the in.-lb-sec system, $g = 386$ in./sec^2.

Once the author was riding up the Mississippi River on a tug that was handling several large barges. The tug owner was going to re-power the ship with a new diesel engine and gearbox and it was the author's job, as the engine manufacturer's representative, to establish the duty cycle of the engines. Did we really want this particular sale for our engine and gearbox?

During the tow the propeller hit a large log, the engine stopped instantly, and the coupling was smoking profusely. Someone jumped over the side and saw that the propeller was bent and a blade was missing, but the verdict was that it was workable. The tow proceeded with the ship shaking so much that coffee would not stay in a cup. Although this did influence the author's decision, he was also interested in knowing the shock load this produced to the coupling and the gearbox.

Some data: rpm $= 200$, $WR^2 = 200,000$ lb-in^2, and $K = 2 \times 10^6$ in.-lb/rad. So

$$T = \frac{\pi(\text{rpm})}{30} \left(\frac{WR^2}{g} \times K \right)^{1/2}$$

$$= \frac{\pi(200)}{30} \left(\frac{200,000}{386} \times 2 \times 10^6 \right)^{1/2}$$

$$= 674,210 \text{ in.-lb}$$

$$= 56,200 \text{ ft-lb}$$

That's a pretty sizable impact, which could have damaged gears and possibly affected the engine. It certainly wrecked the coupling.

2.14   CASE HISTORY: STARTUP TORQUE ON A MOTOR COUPLING

A 7000-hp 1800-rpm motor drives a compressor through a gearbox (Figure 2.24). It is expected to come up to speed in 7 seconds or less, to avoid overheating the motor. A very similar system has a compressor speed of 6000 rpm compressor speed compared to this system's 9000 rpm. Should there be any concern if the same coupling is used?
   The gear is at the motor shaft speed, but the compressor must be converted:

$$800\left(\frac{9000}{1800}\right)^2 = 800(25) = 20{,}000 \text{ lb-ft}^2$$

This is the WR^2 that the motor "thinks" the compressor is. The coupling sees

$$5000 + 20{,}000 = 25{,}000 \text{ lb-ft}^2$$

The torque in the coupling can be calculated from

$$T = \frac{WR^2 \,\Delta\text{rpm}}{307.6t}$$

$$= \frac{(25{,}000)(1800 - 0)}{307.6(7)} = 20{,}900 \text{ lb-ft}$$

For the similar system,

$$WR_S^2 = 5000 + \left(\frac{6000}{1800}\right)^2(800) = 13{,}900 \text{ lb-ft}^2$$

$$T_S = 11{,}610 \text{ lb-ft}$$

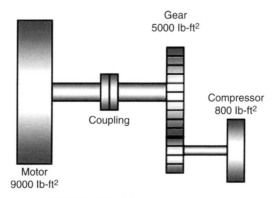

FIGURE 2.24   Motor coupling torque.

The torque is almost double for the new system, so there is some reason for concern. Some idea can be obtained as to whether 20,900 lb-ft is excessive by comparing it with the full-load torque. Since the compressor is unloaded, any torque less than full load should be acceptable.

Using the familiar horsepower equation with T (ft-lb) and N (rpm) yields

$$hp = T\frac{(rpm)}{5252}$$

$$T = 5252 \ hp/rpm = 5252\left(\frac{7000}{1800}\right) = 20,400 \ ft\text{-}lb$$

Since the full load has not been exceeded, the design is acceptable and the same coupling will work. A prudent designer would examine the cost of the coupling that gave the same margin (i.e., 50% of full load). If a change to the next larger size proved feasible, it might be wise to use it.

2.15   CASE HISTORY: FRICTION CLAMPING DUE TO BOLTING

This case was used to understand why the bolts on a motor-driven geared compressor coupling sheared (Figure 2.25). The resisting torque is available from friction clamping of the joint. The clamping force for thread friction from tests is 0.183. Thus,

$$F = \frac{T}{0.183D}$$

$$= \frac{24(12)}{(0.183)(0.375)}$$

$$= 4200 \ lb$$

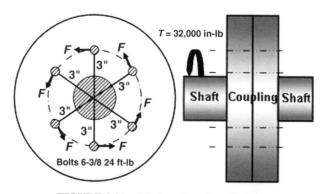

FIGURE 2.25   Friction clamping of a joint.

The torque available from friction is

$$T_{\text{friction}} = (\text{number of bolts})\mu(\text{clamping force})[\text{radius (in.-lb)}]$$
$$= 6(0.1)(4200)(3) = 7560 \text{ in.-lb}$$

The amount of torque reaction that each bolt body takes in shear is the total torque input minus that utilized by friction:

$$T_{\text{shear}} \text{ per bolt} = \frac{32,000 - 7560}{6} = 4073 \text{ in.-lb}$$

This amount of torque is left to shear the bolts since it has not been taken by friction.
    The shear stress per bolt

$$\tau_{\text{bolt}} = \frac{\text{shear load}}{\text{shear area}} = \frac{4073/3}{0.785(0.375^2)}$$

This bolt shear stress of 12,300 lb/in^2 is less than the 60,000 lb/in^2 failure shear stress. The bolt design is adequate and it would take an overload five times the operating load to cause the bolts to fail. This result led the investigators to try to identify the source of a fivefold overload. The cause was a voltage drop on the synchronous motor that can develop air gap torques seven to 10 times the mean torque. This usually occurs when a voltage drop and then a restart attempt occur in too short an interval. A time-delay relay was wired incorrectly and did not function, and the motor restarted too soon after trip-out. Rewiring solved the problem on this new installation.

2.16   CASE HISTORY: FAILURE OF A CONNECTING ROD IN A RACE CAR

This case history illustrates a classical use of fatigue calculations. A small four-cylinder dragster had an engine that could develop 380 hp at 8000 rpm. To increase the strength and reduce the weight, titanium connecting rods were used. The engine failed in operation, with very little evidence left to determine the cause (Figure 2.26). On the piston end of a good rod, a $\frac{1}{8}$-in. oil hole was noted, making this appear to be a weak point in the design. This analysis was performed to see if the rod end was adequate and if this was just a random failure. With a random failure, new rods of the same design would be purchased at considerable cost.
    Figure 2.27 shows a connecting rod geometry and the piston end side with the oil hole. It should be mentioned that the only force on the rod which is of concern is that which develops a tension load in the rod. The firing pressure is not of importance. The controlling components that cause this loading are the piston pin, the piston weight, and any oil that might be in the piston to increase its mass. Inertial loading occurs when the piston rod is reversed in direction from the exhaust stroke to the intake stroke. During this period the inertia of the piston assembly mass is moving it in the direction of the head end, and the connecting rod is moving toward the crankcase end.

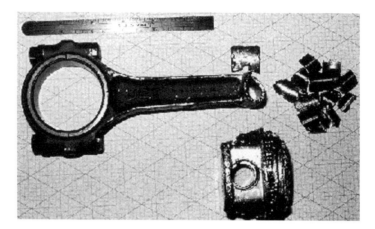

FIGURE 2.26   Connecting rod failure.

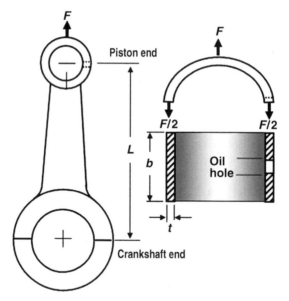

FIGURE 2.27   Connecting rod geometry.

An equation with this loading force F is shown and relates the engine speed, piston weight, connecting rod length, and engine stroke to determine this force.

$$F = \frac{W_{piston}}{g} R\omega^2 \cos\theta + \frac{W_{piston}}{g} R\omega^2 \frac{R}{L} \cos 2\theta$$

where θ is the crank position (deg), $R = 0.154$ ft, $L = 0.56$ ft, $W_{piston} = 1.19$ lb, $\omega = 838$ rad/sec, $g = 32.2$ ft/sec^2, $t = 0.125$ in., and $b = 0.9$ in. The maximum value for F

occurs when the angle is equal to zero and results in the following values for the two terms after the data are inserted:

$$F = 4000 + 1100 \text{ lb}$$

The first term is associated with the crankshaft speed and the second term with twice the crankshaft speed. The total load acting on the top half of the rod is

$$F = 5100 \text{ lb}$$

This is the maximum force, and the minimum force will be zero when the rod cap being analyzed is unloaded. The free-body diagram of the cap is shown, with the force F applied in Figure 2.27. The stress will be determined by considering an axial stress to be acting through the oil hole region:

$$S_{axial} = \frac{kF}{A_{total}}$$

where k is the stress concentration of the hole and A_{total} is the cross section of the rod end (in^2). The stress concentration factor for a $\frac{1}{8}$-in.-diameter hole in a plate in tension is $k = 2.63$, and the total area being acted on by the force F is

$$A_{total} = tb + t(b - \text{oil hole diameter})$$

The maximum and minimum stresses are

$$S_{max} = 64,000 \text{ lb/in}^2 \quad \text{and} \quad S_{min} = 0$$

Material data for titanium 6A1-4V:

Tensile strength	140,000 lb/in^2
Yield strength	108,000 lb/in^2
Endurance strength	58,000 lb/in^2 at 10^6 cycles

Correcting the endurance limit for surface finish (0.8 machined) and axial loading (0.8 axial) results in the corrected endurance limit:

$$S_e = 0.8(0.8)(58,000) = 37,000 \text{ lb/in}^2$$

A quick factor-of-safety check indicates that a fatigue failure is likely, as the value is less than 1.

$$S_{alt} = \tfrac{1}{2}(S_{max} - S_{min})$$
$$S_{avg} = \tfrac{1}{2}(S_{max} + S_{min})$$

$$FS = \frac{1}{kS_{alt}/S_e + S_{avg}/S_y}$$

$$= \frac{1}{(32/37) + (32/108)} = 0.86$$

This is visualized further by a plot of these values. The modified Goodman diagram for this case was shown in Figure 2.10. The diagram clearly indicates that the rod can be expected to fail in the oil hole region in less than 1 million cycles, which it did. Adding material to the rod in the oil hole region, radiusing, and polishing the oil hole were suggested to remedy the problem. From the stress equation it is obvious that doubling the rod thickness t will reduce the stress by half. Plotting this on the diagram shows a substantial improvement.

2.17 BOLTING

Bolting is one of the most common methods for attaching or mounting mechanical parts. Broken or loose bolts on machinery, pressure vessels, piping, and structures can result in substantial losses in both production and repair costs, together with safety concerns. There are many important areas in an analysis of bolting, but in this section three areas are reviewed:

- The holding capacity of the bolt
- The torque required to develop the holding capacity
- The stress induced in the bolt body

2.17.1 Holding Capacity

For low-carbon-steel bolts, a useful equation that relates the tightening torque T (ft-lb), the load induced in the bolt F (lb), and the nominal bolt diameter D (in.) is

$$T = \frac{fFD}{12} \qquad \text{ft-lb}$$

The holding capacity due to torquing the bolt is

$$F = \frac{12T}{fD} \qquad \text{lb}$$

The friction coefficients for various lubricants are listed in Table 2.9.

2.17.2 Limiting Torque

In tightening up bolts or overtightening bolts, the bolt can fail in the threads or pull the threads out of a threaded hole. The latter is common if an upgraded

TABLE 2.9   Bolt Friction Coefficients

Type of Thread Lubricant	f
Teflon coating	0.07
Never Seize	0.12
MolyKote	0.12
Colloidal copper	0.12
Heavy oil	0.12
Dry: cadmium plating	0.15
Dry: steel	0.20
Dry: stainless steel	0.30

high-strength bolt is used in an old piece of machinery such as an old casting. The bolt might be high strength and have the ability to accept high torque, but the mating piece may not. In cast iron equipment, a rule of thumb sometimes used is to maintain the bolt stress at 20,000 lb/in² or less and use a bolt engagement greater than $1\frac{1}{2}$ diameters.

An equation that can be used to relate the torque on the bolt to the nominal stress in the bolt is

$$T = \frac{SfDA}{12} \quad \text{ft-lb}$$

where S is the nominal induced stress at the thread root (lb/in²) and A is the root area (in²). When the exact thread root area is unavailable, a rough approximation based on using 80% of the nominal diameter for calculating the root diameter is

$$T = \frac{SfD^3}{24} \quad \text{ft-lb}$$

$$S = \frac{24T}{fD^3} \quad \text{lb/in}^2$$

It is important to realize that just using a different type of lubricant on threads (i.e., changing f) can increase the bolt load and the stress in the bolt considerably.

For example, consider that a bolt has had no lubrication, $f = 0.2$, and because such bolts are rusting, it is decided to use Teflon-coated bolts, $f = 0.07$. Using the same type of bolt and torque, we obtain

$$\frac{S_{f=0.07}}{S_{f=0.2}} = \frac{24T/0.07D^3}{24T/0.2D^3}$$

$$= \frac{0.2}{0.07} = 2.86$$

TABLE 2.10 Bolt Tensile and Yield Strength

Bolt Material	Tensile Strength (lb/in^2)	Yield Strength (lb/in^2)
Low-carbon steel	64,000	52,000
Medium-carbon steel	100,000 to 150,000	80,000 to 120,000
High-carbon alloys	150,000 to 200,000	120,000 to 160,000

Source: Ref. [8].

TABLE 2.11 Bolt Fraction of Yield

Condition	Fraction of Yield Strength
Nongasket, noncritical static loads	0.25
Gasketed ASME code, history of problems	0.50
Nongasket, leaks, loosens even with torque control	0.75

Source: Ref. [9].

This means that if the bolt had originally been stressed to 50,000 lb/in^2 when torqued, and maintaining that torque, the bolt stress would now be

$$S = 2.86(50,000) = 143,000 \text{ lb/in}^2$$

This could overstress the bolt and overload the gaskets or flanges. This needs to be considered when using low-friction coatings on bolts and studs.

The tensile and yield strength of bolts have a wide range. Some common bolt materials are listed in Table 2.10. Some yielding guidelines when considering the allowable bolt stresses to use are provided in Table 2.11. When dealing with high-carbon alloys, the applications are usually quite critical. It is wise to use specialized vendors or consultants since the alloys can be susceptible to various corrosion mechanisms, such as stress corrosion cracking or hydrogen embrittlement.

Since S is only the nominal axial stress, no consideration has been given to stress concentration, combined loading, manufacturing inaccuracies, or yielding of the bolt. When these unknowns are considered, 30,000-lb/in^2 induced stress seems appropriate as a guideline. It represents a starting point. Consideration of flange or part distortion, mating part thread strength, and gasket tightness all have to be considered, and for some cases even this stress may result in too high or too low a bolt load.

2.17.3 Bolt Elongation and Relaxation

It is important to know how much a bolt elongates when it is installed. With very little elongation of a bolt, a slight relaxation of the material under the bolt, in the threads or the gasket, can cause the bolt to lose its preload. For example, if a bolt after assembly stretches 0.001 in. and the surface under the bolt head is roughly machined, the bolt can eventually be expected to flatten the surface. With no stretch left, the bolt can become loose. The equation for the stretch δ of a bolt or stud with $f = 0.2$ is

$$\delta = \frac{240TL}{\pi D^3 E} \quad \text{in.}$$

where T is in ft-lb, D is in inches, L is the free length (in.), and $E = 30 \times 10^6$ lb/in.^{2}.

The expected relaxation in the threads due to surface embedment of at least the following value can be expected from tests:

$$\text{embedment } \varepsilon = 0.1\delta + 0.001 \text{ in.}$$

For multiple interfaces where embedment can occur, multiply ε by the number of interfaces. A bolt holding one component is one interface, holding two is two, and so on. This relaxation can be greatly reduced by retightening the bolting after several hours of operation. Remember that this embedment is a loss of stretch and thus bolt load:

$$\text{remaining stretch after relaxation} = \delta - \varepsilon$$

$$\text{remaining bolt load after embedment} = \frac{\delta - \varepsilon}{\delta} F_b$$

2.17.4 Torquing Methods

Correct and accurate torquing of a bolt is important and all methods have an associated accuracy. Torque wrench accuracy can be $\pm 15\%$ and depends on the mechanic's expertise and wrench calibrations. Considerable torque can be applied to bolts, and typical methods used in machinery repair in industry are shown in Table 2.12. This information shows that it is quite easy to overtorque a bolt or stud if proper torque measuring is not adhered to. These data can be useful in troubleshooting a failure. If mechanics use a $\frac{1}{2}$-in. drive impact to torque a bolt that requires 1500 ft-lb, it is evident that the bolt was probably undertorqued.

TABLE 2.12 Common Torquing Methods

Torquing Method	Torque Possibility (ft-lb)
Mechanic with a 3-ft wrench	Up to 1000
$\frac{1}{2}$-in. drive impact wrench	Up to 500
1-in.-drive impact wrench	Up to 1000
Hammer or slug wrench	Over 3000
Torque multiplier or hydraulics	Over 3000

An approximate method to determine to what value a bolt or stud was tightened is to check the breakaway torque. Basically, the torqued bolt is untorqued and the torque to "break loose" is noted. This is approximately 80% of the tightening torque and is a method to determine if bolts have remained uniformly tight. Obviously, there will be considerable error with rusted bolts or bolts with galled threads.

2.17.5 Fatigue of Bolts

Once a bolt is torqued to a given value and is therefore prestressed, what effect does an external load acting on the bolt have? This is important since it is this load that can cause bolt failure, especially with cyclic loading applied. Cylinder head and piston rod bolts are examples of such loading.

Bolts, studs, and parts that clamp together can be represented as springs. Each compresses or elongates to some degree when a preload or external load is applied. As long as the preload F in the bolt is greater than the external load F_E applied, the following equation is valid:

$$F_b = F + \frac{K_b}{K_p + K_b} F_E$$

where

$$F_b = \text{total load in bolt (lb)}$$

$$F = \text{load in bolt due to torque (lb)}$$

$$F_E = \text{externally applied load (lb)}$$

$$K_b = \text{spring constant bolt (lb/in.)}$$

$$K_p = \text{spring constant part (lb/in.)}$$

Notice in the equation for F_b that when the bolt is very flexible relative to the part (i.e., K_b is much smaller than K_p), the external load F_E will add only a very little to the bolt load. Almost all the load is due to F, the torque-induced preload. This means that the bolt axial stress will see very little additional stress due to the external load.

The second extreme case is that where the bolt is very rigid and the part is very flexible. Here, K_b is much larger than K_p and the term approaches 1. The full effect of the external load F_E will be added to the preload. This is the worst condition, and if the value of K_p is unknown, the prudent engineer will design for this case. Bolt fatigue failures occur somewhere between these two extremes.

The spring constant K_b for the bolt is simply the load divided by the elongation:

$$K_b = \frac{A_b E}{L} \quad \text{lb/in.}$$

TABLE 2.13   Spring Constant Ratios

Case Experienced	Value of $\dfrac{K_b}{(K_p + K_b)}$
Metal to metal with long bolts	0
Steel bolt holding two steel plates	0.2
Hard copper gasket with long bolts	0.25
Soft copper gasket with long bolts	0.5
Two plates and an asbestos gasket	0.6
Soft packing through bolts	0.75
If you have no idea and want a safe value	1.0

$A_b = \pi D^2/4 \text{ in}^2$, L is the free length to elongate (in.), and $E = 30 \times 10^6 \text{ lb/in}^2$. The spring constant of the part K_p is usually difficult to determine, and in most cases only approximations are possible. When possible, tests can be conducted by applying a known load to a part and accurately measuring the displacement under the load, within the elastic range. This load, divided by the displacement, is the spring constant. This is especially useful when stacking of parts or gasketed parts is considered.

For a part with a circular area A_p undergoing compression under the bolt head, an order-of-magnitude estimation can be established by using the equation

$$K_p = \frac{\pi D_p^2 E}{4L} \qquad \text{lb/in.}$$

Here D_p is the zone under the bolt of diameter D_p, which undergoes displacement over its length L. As a rough approximation, use D_p equal to three times the bolt diameter. Ratios experienced in industry are listed in Table 2.13.

Consider a split-case horizontal pump with a flat gasket. It has a pressure load of 4000 lb per bolt and a bolt preload of 18,000 lb. How much load will the bolt see, and how much will it elongate?

$$F_b = F + \frac{K_b}{K_p + K_b} F_E$$
$$= 18,000 + 0.6(4000) = 20,400 \text{ lb}$$

How much will a $\frac{3}{4}$-in. bolt 10 in. long elongate due to the pressure?

$$\delta = \frac{PL}{AE}$$
$$= \frac{0.6(4000)10}{0.785(0.75^2)(30 \times 10^6)}$$
$$= 0.0018 \text{ in.}$$

Gasket recovery will have to adjust for this or the joint will leak. In a solid clamped assembly the preload bolt stress is reduced at a rate of 30,000 lb/in² per 0.001 in. loss

TABLE 2.14   Stripping Strength of Threads

Bolt Size	Steel: Shear = 50 ksi	Aluminum: Shear = 40 ksi	Cast Iron: Shear = 30 ksi
$\frac{1}{4}$ to 1 in.−NC	0.25D	0.35D	0.45D

of grip, so maintaining grip is important. Some sources recommend that the preload F be at least twice as much as the external load F_E.

2.17.6   Stripping Strength of Threads

Bolts tightened into threaded holes can strip out if overtightened. Although a bolt manufacturer of high-alloy bolts may say that the bolt can be tightened to 100,000 lb/in², the material it threads into, such as cast iron, could break or strip out. Table 2.14 indicates the engagement at which the threads will strip with a 50,000-lb/in² bolt stress and the material shown. These are test data [10] that provide old but still valid information. As developed in a case history in Section 10.3.3, this can be extended to the strip-out strength of larger-diameter internal threads by using the following equation:

$$L = 0.225 \frac{\sigma_{\text{shear stud}}}{\sigma_{\text{shear part}}}$$

For design purposes, 1.5D is normally used to ensure that no stripping occurs, where D is the nominal bolt diameter (in.). As an example, assume that it is required to calculate the thread engagement for a $\frac{1}{2}$-in. bolt to withstand a 50-ksi (50,000-lb/in²) bolt load in cast iron. From Table 2.14, 0.45D or an engagement length of 0.45 $(\frac{1}{2}) = 0.23$ in. minimum is determined as being the length the material will strip. Engagement should be greater than this.

2.17.7   Case History: Power Head Gasket Leak

A 2700-hp piston engine compressor periodically develops a leak under one of the eight cylinder heads. When this occurs, the engine must be shut down and the cylinder head bolts tightened. Production losses occur during this period, as the equipment is unavailable. For safety reasons, tightening the bolts when the engine is operating is unacceptable. The problem to be solved is what can be done to eliminate the leaking.

Figure 2.28 illustrates how the power head is held in place. It is not an exceptionally well-thought-out design, since there is quite a stack-up of clamped items. Since the design is fixed and only bolt or gasket changes are possible, the bolt design is reviewed first. Tightening the bolts appears to solve the problem if the gasket doesn't blow out. The question is: How much tightening is required?

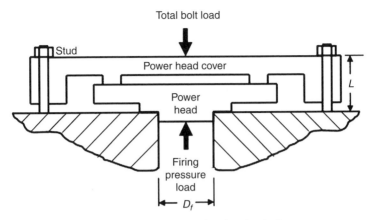

FIGURE 2.28   Power head gasket leak.

Engine data:

Bolts	10 $1\frac{1}{2}$ in. in diameter, 15 in. long, 105,000 lb/in²
Torque	1-in. impact wrench
Firing pressure	1500 lb/in² at detonation
Power cylinder diameter	17 in.
Dry threads	$f = 0.2$

The force acting on the bolt due to the firing pressure trying to lift the head off is

$$F_f = P_f A_f = P_f \frac{\pi D_f^2}{4}$$
$$= 1500 \, \frac{\pi(17^2)}{4}$$
$$= 341,000 \text{ lb acting on the bolts}$$

The clamping force of the bolts holding the head down is

$$F = \frac{60T}{D}$$

Assume a torque of 1000 ft-lb due to the 1-in. impact wrench:

$$F = \frac{60(1000)}{1.5} = 40,000 \text{ lb/bolt}$$

Since there are 10 bolts, the total load holding the head down is 400,000 lb. Without going too much further, a problem is obvious. The load on the head under detonation is 341,000 lb, and the clamping load due to the bolts is 400,000 lb. This is not enough of a design margin, as the type of torque wrench used is very inaccurate, and the analysis is only approximate.

An accurate hydraulic wrench with torque readout was available that had a range to 3000 ft-lb. It was decided to see if 1500 ft-lb would provide adequate clamping and not overstress the bolt or cause pull-out from the cast iron block. The new clamping force calculated by a ratio of the torque increased the design margin from 1.17 to 1.76, which is much better.

$$\frac{400,000(1500)}{1000} = 600,000 \text{ lb}$$

The bolt stress at the new condition is

$$S = \frac{120T}{D^3} = \frac{120(1500)}{1.5^3}$$
$$= 53,300 \text{ lb/in}^2$$

This bolt stress is acceptable, and since the bolt had a two-diameter engagement into the cast iron block, thread pull-out was not a problem.

As a final check, bolt stretch was calculated to verify that relaxation would not be a concern. Stretch of over 0.010 in. was felt from experience to be acceptable.

$$\delta = \frac{240TL}{\pi D^3 E}$$
$$= \frac{240(1500)(15)}{\pi(1.5^3)(30 \times 10^6)}$$
$$= 0.017 \text{ in.}$$

The outcome was successful and no future leaks occurred when the bolts were maintained at 1500 ft-lb. The analysis could have been much more exhaustive. Items such as system flexibility, gasket compression, thermal effects, and more accurate firing pressure data could have been analyzed. In this case the key point was that field tightening had helped. With this information all that was left for the engineer to do was to determine how much tightening was required.

At a later date, tests were conducted to determine the spring constant of the clamped parts under the bolt and to determine the cyclic stress in the bolt. The results were

$$K_b = 14 \times 10^6 \text{ lb/in.}$$
$$K_p = 83 \times 10^6 \text{ lb/in.}$$

The bolt load is

$$F_b = F + \frac{K_b}{K_p + K_b} F_E$$

$$= \frac{600,000}{10} + \frac{14}{83 + 14}\left(\frac{341,000}{10}\right)$$

$$= 60,000 + 0.14(34,000)$$

$$= 60,000 + 4900 \text{ lb}$$

In terms of stress,

$$S = 53,300 + 4400 \text{ lb/in}^2$$

On a Goodman diagram this was not a problem. The author uses a table such as Table 2.15 to document failures.

This can all be summarized by suggesting that three bolting equations be remembered. They are based on stressing alloy steel studs or bolts to 50,000 lb/in². Torque bolts to

$$T_{\text{ft-lb}} = 420D^3_{\text{nominal diameter (in.)}}$$

TABLE 2.15   Documentation of Bolt Failures

Application	Bolt diameter–length	Torque (ft-lb)	Relaxation Stretch	Preload Alternating	History
Eccentric weight vibrating conveyor	$\frac{3}{4}$–10	200	$\frac{0.002}{0.012}$	14.2	OK
Agitator paddle	1–3	150	$\frac{0.001}{0.001}$	1.1	Fail bending, find loose
Agitator paddle	1–3	500	$\frac{0.001}{0.004}$	4.3	OK
Cutter blade reprocessor	$\frac{3}{4}$–2	275	$\frac{0.003}{0.003}$	4.9	Bolts loose, raised to 450 ft-lb; OK
Cyclic tension test bolt	$\frac{3}{8}$–1	7	Loose	0	Fail 6000 cycles
Cyclic tension test bolt	$\frac{3}{8}$–1	40	Tight	10	No failure, 10^7 cycles
Main bearing cap	$\frac{5}{8}$–5	5	Loose	0	Not tightened; failed in fatigue

Clamping load should be more than operating load; some say two times:

$$F_{lb} = 39,250 D^2_{\text{nominal diameter (in.)}}$$

Bolt stretch should be greater than 0.002 in.:

$$\delta = 0.0017 L_{\text{free length (in.)}}$$

For bolt nominal stresses of less than 50,000 lb/ in², just multiply the equations by

$$\frac{\text{new stress (lb/in}^2)}{50,000 \, \text{lb/in}^2}$$

2.18   BALL AND ROLLER BEARING LIFE ESTIMATES

In Section 2.17 the loads on a bearing were determined and from this it was shown that the bearing life was greatly reduced with increases in the load due to imbalance. On machinery, the B_{10} *life* is used as a major criterion on the acceptability of a ball or roller bearing. The B_{10} life is a statistical term that defines a given group of approximately identical bearings operating under equal conditions of load and speed. It means that 90% of the bearings can be expected to meet this fatigue life before pitting and spalling occur. The arithmetic average life is seven times the B_{10} life.

   To determine a ball or cylindrical roller bearing life, the following equation can be used as an approximation:

$$B_{10} \text{life} = \frac{1.93}{\text{rpm}} \left(\frac{C}{P}\right)^R \qquad \text{years}$$

$$R = \begin{cases} 3 & \text{for ball bearings} \\ \dfrac{10}{3} & \text{for roller bearings} \end{cases}$$

In this equation, P is the equivalent radial load and C is the basic load rating of a bearing in pounds. C is supplied by the bearing manufacturer for each bearing. The formula by itself reveals some significant information:

1. Halving the rpm doubles the life.
2. Halving the load P increases the life eight to 10 times.
3. Bearings with large C values will have longer lives.

   For ball bearings, P is sometimes not a pure radial load but also has to support some thrust load. In such cases, the equivalent load can be approximated by using the larger of the following:

$$P = 0.56 F_R + 2.3 F_A$$

$$P = F_R$$

It is always best to use the bearing manufacturer's engineering catalog, but the method above will produce answers on the conservative side, that is, larger P values.

2.18.1   Case History: Bearing Life of a Shaft Support

An additional bearing was proposed by an aftermarket parts supplier. The bearing was to add a support point to a very flexible vertical shaft (Figure 2.29). A ball bearing was specified based on the available space; however, it looked too light for this service and the engineer decided to do a quick life calculation.

Proposed bearing and calculated data:

Bore	110 mm
OD	170 mm
Width	28 mm
C	14,200 lb
rpm	700
Axial load	$F_a = 1000$ lb
Radial load	$F_R = 1600$ lb

$$P = 0.56F_R + 2.3F_A$$

$$= 0.56(1600) + 2.3(1000)$$

$$= 3,200 \text{ lb}$$

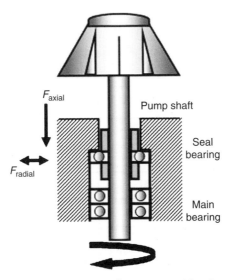

FIGURE 2.29   Axial flow pump seal bearing.

$$B_{10} \text{ life} = \frac{1.93}{\text{rpm}} \left(\frac{C}{P} \right)^R$$

$$= \frac{1.93}{700} \left(\frac{14,200}{3200} \right)^3$$

$$= \frac{1}{4} \text{ year}$$

This is not adequate; the B_{10} life designed for this machine should be longer than three years.

If the equation is rearranged, the C rating required can be determined for a B_{10} life of three years and bearing catalogs used to locate a bearing with this C value.

$$C = P \left[\frac{(\text{rpm})B_{10}}{1.93} \right]^{1/3}$$

$$= 3200 \left[\frac{700(3)}{1.93} \right]^{1/3}$$

$$= 32,913 \text{ lb}$$

A bearing was selected with this C value and the correct dimensions, and the information was relayed to the bearing manufacturer.

The importance of this quick check should not be underestimated. This piece of machinery was remotely located, and a bearing failure would have resulted in a very costly wreck.

2.18.2 Coupling Offset and Bearing Life

When a coupling manufacturer's catalog states that the maximum allowable offset is 50 mils for a gear coupling, the user is cautioned. The ball and roller bearing life can be reduced if operated at the extreme. As misalignment is increased, so is the load produced. The higher loads result in a shorter bearing life. In Figure 2.30, bearing

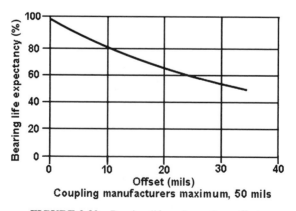

FIGURE 2.30 Bearing life and coupling offset.

life and offset do not consider angular misalignment [11]. Note from the figure that 50% of the bearing life is lost and the coupling is not yet at the manufacturer's maximum value. This demonstrates the importance of keeping angular misalignment to a minimum. For high speeds, less than 1 mil of offset or 1 mil/in. angular misalignment is generally used.

2.19   HYDRODYNAMIC BEARINGS

Antifriction bearings such as rolling element bearings are not the only important bearings in large machinery. Journal bearings are used extensively in high-performance machines such as:

- Centrifugal compressors
- Steam and gas turbines
- Motors
- Reciprocating compressors
- Reciprocating engines (automotive and industrial)
- Gearboxes

Journal bearings should not be confused with bushing-type bearings, which are impregnated with oil and not supported by hydrodynamic film.

Hydrodynamic bearings are journal or pad-type bearings that develop a pressure film due to the shaft and bearing geometry and motion. This pressure profile supports the journal load on a thin film. To better explain the parameters, let's review Figure 2.31 and the basic equation on this support mechanism. Figure 2.31 is greatly exaggerated for clarity; the journal or shaft does not climb this high up the side of the bearing

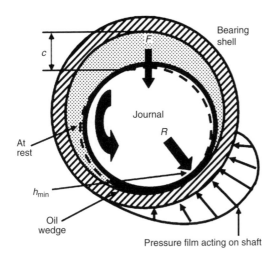

FIGURE 2.31   Hydrodynamic bearing support mechanism.

shell. In the sketch are shown the supporting oil film pressure distribution, the load F, and the oil wedge. The oil is dragged into the gap on the surface of the journal and develops this pressure. The minimum oil film seen is shown as h_{min}, and the diametral clearance at rest is c. This clearance is measured during assembly and is different from the running clearance, which varies around the shaft during operation.

The following simple equation [12] for a 360° journal bearing with an L/D value of 1 illustrates how some of the parameters affect the minimum oil film thickness:

$$h_{min} = \frac{1}{2}c - \frac{3F}{\mu L(\text{rpm})}\left(\frac{c}{D}\right)^3$$

where

$$h_{min} = \text{minimum oil film thickness (in.)}$$

$$c = \text{diametral clearance (in.)}$$

$$F = \text{load (lb)}$$

$$\mu = \text{oil viscosity (lb-sec/in}^2\text{)}$$

$$L = \text{bearing length (in.)}$$

$$\text{rpm} = \text{Shaft speed}$$

$$D = \text{journal diameter (in.)}$$

For a typical engine bearing, if the load F is doubled, the minimum film is reduced by 60%. The film can support a heavy load, and as the load gets heavier, the film just gets thinner. Also, by reducing the viscosity by a factor of 2, because of incorrect viscosity or a higher temperature, the film is reduced by 30%. If both are reduced, the film can be reduced by 90%.

A case history together with additional information on the use of this equation is presented in Section 10.6.3. The design of these types of bearings is critical. Some typical values the author has experienced include minimum films of about 0.0005 in. To pass within this minimum film, surface finishes need to be in the range 16 to 32 μin. rms. Also, when the maximum bearing temperatures get to 250°F, the tin-based babbitt layer tensile strength is reduced to half its room-temperature value. Too much higher and localized melting of the babbitt can start to occur.

Analysis of these types of bearing can get complicated very quickly, due to the dynamics of the system. Bearing types used in turbo machinery are:

- Double-acting tilting pad bearings for thrust
- Tilting shoe–type journal bearings
- Journal bearings on reciprocating machines

Tilting bearings can be self-leveling; however, they all work on a similar principle of support on a film. The differences in designs are generally used to "tune" the system in terms of damping and spring constants to handle the many instabilities that can arise in complex high-speed, high-horsepower turbo machines.

Journal bearings for reciprocating machines and almost all automobile engines are as shown in Figure 2.31. All the oil pump does is to keep the bearing filled with oil, which then leaks back to the sump from the end clearance. Lose the pump and you lose the oil supply so that no film can develop and metal-to-metal contact occurs.

Almost all wear takes place during startup or shutdown as the shaft is at low speeds. With clean oil, the correct oil, and smooth operation, these types of bearings can last the life of the machine. Unfortunately, this doesn't always occur.

2.19.1 Shell and Pad Failures

Many failure mechanisms occur in the types of bearings we have been discussing. The following failures are those experienced by the author over and over again. Sketches such as those shown in Figures 2.32 to 2.35 are used. The shell-type bearings have been visually flattened to show the damage. What an observer would see looking down on a shell half is shown in Figure 2.32.

Edge loading is usually due to shaft deflection. On thrust pads it can be caused by the pad not being free to pivot and self-align. On internal combustion engines such as diesel or gasoline, it can be alignment with the crankshaft and frame or an adjacent bearing that has failed.

When there are hard particles passing between the bearing half and the journal, the particles need to be smaller than the minimum oil film thickness. When they are not smaller, they will gouge out a path in the soft babbitt as the shaft rotates. Large

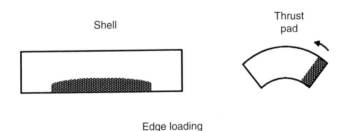

Edge loading

FIGURE 2.32 Bearing edge loading damage.

Dirt scratches

FIGURE 2.33 Bearing hard particle damage.

Normal and heavy wear

FIGURE 2.34   Bearing normal and heavy wear.

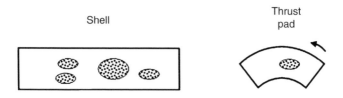

Fatigue failure or
disbonding

FIGURE 2.35   Bearing fatigue failure or disbonding.

particles may embed themselves in the babbitt and act as a grinding tool on the shaft
or journal, scoring it. A clean oil system does not have this problem. Check the size
of the filters and make sure that they are not bypassing unfiltered oil. An oil analysis can tell if the material has been generated internally; that is, it might be from a
failing part.

In the author's experience, the difference between the normal wear of a journal
bearing and heavy wear is just a matter of degree, which is why they are shown
together in Figure 2.34. The wear tends to be uniform, and in trimetallic bearings, a
couple of layers may be worn away. When melting, spalling, pitting, and discol-
oration are not associated, the wear is usually normal after many years of operation.
Automobile engines with over 100,000 miles can look like this when they have
been well cared for. Usually, some scratching is also present from carbon deposits
passing through. Reciprocating aircraft engines also have this appearance after
3000 hours of operation. Thrust pads are usually discolored, with the leading edges
slightly melted or the entire pad smeared with babbitt when they have been loaded
excessively.

Fatigue failures and a poor bond look similar. Chunks of babbitt may be missing.
In fatigue, pieces are usually still attached to the backing, but with a poor bond are
usually free from the backing. This is not always the case, however; a metallurgical
examination will usually determine which occurred. This is important to know,
because disbonding is a quality control or corrosion issue. Fatigue is due to continual
pounding of the babbitt surface by the operating loads. Internal combustion engines
usually experience this type of failure when they are subjected to heavy detonation.

This is discussed in Section 6.14. What happens is that the babbitt cracks, and like the debris in a road pothole, it keeps breaking up into smaller pieces. When the pieces are smaller than the minimum oil film, they are washed out and leave a rather smooth cavity in the bearing. Eliminating the cause of the pounding or going to a thinner babbitt or higher-strength material is usually required to solve the problem. Thinner babbitt can withstand pounding better, but at a sacrifice to embeddability.

2.20 GEARS

Gear units are used to speed up or slow down driven machines such as compressors, extruders, pumps, and drivetrains. The drivers are usually steam or gas turbines or electric motors. Most gear design is done by the manufacturer of the gearbox unit following the recommendations of the American Gear Manufacturers Association (AGMA).

Engineers usually have a need to know the details of gear design when one of the following occurs:

- A gear unit fails and broken teeth, pitting, or scoring is visible. Was the gear overloaded or was there some other cause?
- A new machine is being ordered and the engineer wants to know if it is adequate. The design may be unique or the engineer may be unfamiliar with the manufacturer's engineering department.
- The throughput of a production process is being increased and a higher-input-horsepower driver is being installed. The gear manufacturer says that the gearbox can handle the higher throughput, but the engineer would like to verify this.

2.20.1 Gear Acceptability Calculations

Three calculations will indicate the acceptability of a gear tooth design: the bending strength, pitting resistance, and scoring index. Several methods to calculate each are available, some more accurate than others. The equations presented here are simple to use and will generally result in conservative values; that is, the gear will be stronger than the calculations indicate. The first two calculation methods are from the American Petroleum Institute Standard API 613. The AGMA calculation method, AGMA 218, should be used on marginal designs or when fine tuning a design. In the following calculation methods, additional simplifications have been used for these quick checks.

The *bending strength number* (BSN) of gear teeth relates to the stress at the root of the gear and is directly related to the torque applied to the shaft:

$$BSN = \frac{680,000\,(hp)n}{(rpm)d^2F}$$

where

$$hp = \text{horsepower to gear}$$

$$n = \text{number of gear teeth}$$

$$rpm = \text{gear speed}$$

$$d = \text{pitch diameter of pinion (in.)}$$

$$F = \text{face width (in.)}$$

When a 1:1 gearbox is used, the horsepower is divided between the main shaft and the auxiliary shaft. If the diametral pitch, P, is provided by the manufacturer, the pitch diameter $d = n/P$.

Calculations for the BSN should result in values that do not exceed the following:

For through hardened and nitrided teeth:

30,000 for BHN 360

20,000 for BHN 220

For carburized teeth:

40,000 for Rockwell C (Rc) 60

35,000 for Rc 53

These values are based on a safety factor of 1.5 and a J value of 0.5 as defined in API 613.

The *tooth pitting index* (TPI), calculation determines if breakdown of the tooth surface due to fatigue is a problem. This represents a simplified version of a relatively complex calculation which considers such factors as strength of materials as related to pitting resistance as well as geometric factors associated with the teeth.

$$TPI = \frac{126,000(\text{hp})(R + 1)}{F \times (\text{rpm})d^2 R}$$

R, the gear ratio, is defined as the number of teeth in the larger gear divided by the number of teeth in the smaller pinion. For a safety factor of 1.5, the TPI calculated should be less than the values shown in Table 2.16.

The *scoring index* calculation determines if there is a chance that the welding and tearing effects of high tooth loading will occur. There are several methods for calculating the potential for scoring, one of the simplest being to apply the *Dudley scoring index* (DSI). It is presented here because of its simplicity and because it seems

TABLE 2.16   Tooth Pitting Index Limiting Values

Condition	Hardness	Limiting Value
Carburized teeth	Rc 55–60	Less than 295
Gas-nitrided steels	Rc 50–60	200–225
Through hardened	BHN 360	Less than 185
	BHN 300	Less than 135
	BHN 250	Less than 100

to work. Carburized hardened gears above 450 BHN with pitch line velocities above 15,000 ft/min are most susceptible to scoring. The equation assumes 150 SSU oil, 120°F, oil temperature, and a carburized tooth surface. A DSI value below 14,000 indicates a low probability of scoring, whereas values above 18,000 indicate a high probability. Unless values are quite low, further investigation into the possibility of scoring is advisable.

When reviewing a gear's operation by examining the teeth mating pattern, the usefulness of the calculations becomes apparent. For example, if only 20% of a gear tooth is in contact and on the edge of the gear, the effective face width F is reduced by 80% and all the index values increase by a factor of 5. Poor alignment of gears due to a failed bearing usually results in this condition, causing well-designed gears to fail in bending or pitting. On helical and spur gears it is important that the teeth be checked for contact pattern. Even contact across 80 to 90% of the tooth face is desirable. Heavy loading at the end or top of the teeth is unacceptable.

The reason that 80% contact is acceptable is that the gear is expected to spread out under loaded conditions without heavy loading occurring on the ends of the teeth. Coating several of the gear teeth with prussian blue dye greatly aids in viewing the contact pattern. The dye should be painted on thin so as not to produce a false contact pattern. When poor contact occurs on a new gearbox design or an old design that always had pitting, the following design guide can be useful:

$$\frac{\text{span between bearings}}{\text{pinion pitch diameter}} < 2.5$$

This check roughly calculates the deflection of the pinion on the gear (Figure 2.36), which should be kept less than 0.001 in., as excessive deflection alters the contact pattern.

A useful equation [13] that calculates the tooth separation deflection δ due to bending and torque is

$$\delta = \frac{\text{hp}}{1000D^3(\text{rpm})}\left(\frac{1.1L_b^4}{D^2L} + 10L\right) \quad \text{in.}$$

where D is the pitch diameter (in.), L the tooth width (in.), and L_b is the bearing span (in.). The first term, $1.1L_b^4/D^2L$, represents the lateral bending, and the second term, $10L$, represents the torsional twist effect on the separation.

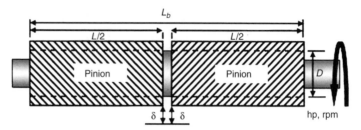

FIGURE 2.36   Bending of a pinion due to load.

When the shaft is hollow and the bore diameter is d, the deflection is

$$\delta_{\text{hollow}} = \delta \frac{1}{1 - (d/D)^4} \quad \text{in.}$$

Here's an example of solid shaft pinion deflections:

$$\delta = \frac{hp}{1000 D^3 (\text{rpm})} \left(\frac{1.1 L_b^4}{D^2 L} + 10L \right)$$

$$= \frac{5228}{1000(5.12^3)(11{,}942)} \left[\frac{1.1(22.25^4)}{(5.12)^2(6.7)} + 10(6.7) \right]$$

$$= 0.0052 \, \text{in.}$$

This seems an unacceptable deflection for this gear. Notice that the shaft between the gears is assumed to be the same as the pitch diameter D. The bearing shaft is assumed to have no contribution to the deflection. This is not always the case.

In gearboxes that have been installed for many years in the same service, it has been the author's experience that the failures analyzed have never been caused from the gears having obtained their design life and having just worn out. That also goes for the bearings in the gearboxes. Many examples of gear faces and bearing pitting and spalling have been observed, as would be expected from end-of-life fatigue failures. In all cases, however, these were secondary effects. With the load conditions as designed, the oil kept clean and at the correct temperature, and startup and operating conditions as designed, gearboxes and bearings will last a long time. The failures occur most often after re-rating to higher loadings or are due to continued abnormal operating or startup conditions. Outside effects such as misalignment of the gearbox between the driver or driven machines can also result in broken shafting, failed bearings, and failed gearing. Figure 2.37 shows a gear tooth with an end load distribution that can break teeth. Normally, the full face width distributes the load, but with end loading this is not so and the end break shown can occur. This was caused in this case by a bearing failure, which resulted in abnormal loading on the end of the tooth. The bearing failure was due to excessive misalignment, which caused high loads on the gearbox output shaft. The cause of the misalignment, lockup of a splined coupling, is examined in Section 10.4.2.

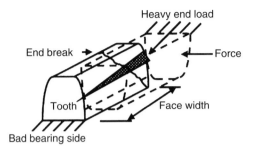

Heavy end load

End break → Force

Tooth   Face width

Bad bearing side

FIGURE 2.37   Gear tooth failure due to a series of events.

Gearbox failures have a way of masking the true cause of a failure. It usually takes a considerable amount of detective work by experienced engineers, analytical modeling, and metallurgical examination to determine the most likely cause. Notice that the failure is opposite the bad bearing side. This is because the gears spread apart at the worn-out bearing end, and the high load was on the opposite end. In most cases the extra time and cost to do a thorough analysis is insignificant compared to the production downtime costs associated with a repeat failure.

2.20.2   Case History: Uprate Acceptability of a Gear Unit

In this case the production rate of a process was being reviewed for a possible rate increase. One question that needed to be answered was if the gear unit could handle the increased horsepower requirements. Over the years this gear unit had undergone several failures for various causes, so it was suspect. The input horsepower to the gear unit was 600, with the possibility of being increased to 800.

Additional data on the gear unit are as follows:

$$d = 5.5 \text{ in.} \qquad R = 1$$
$$F = 10 \text{ in.} \qquad n = 24 \text{ teeth} \qquad \text{hp} = 300 \text{ each gear}$$
$$\text{rpm} = 305 \qquad \text{gear through-hardened} \qquad \text{hardness} = 360 \text{ BHN}$$

In the simplified calculations we first review the gears at the current design conditions, since failures at these conditions have occurred. The gear pair is shown in Figure 2.38 in relation to the input and output shafts.

Calculating the bending strength number:

$$
\begin{aligned}
\text{BSN} &= \frac{680,000(\text{hp})n}{(\text{rpm})d^2F} \\[6pt]
&= \frac{680,000(300)(24)}{305(5.5^2)(10)} \\[6pt]
&= 52,400
\end{aligned}
$$

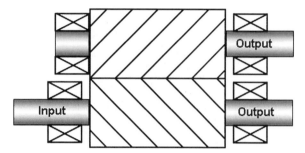

FIGURE 2.38 Gear pair uprate calculations.

Calculating the tooth pitting index:

$$\text{TPI} = \frac{126,000(\text{hp})(R+1)}{F(\text{rpm})d^2R}$$

$$= \frac{126,000(300)(1+1)}{10(305)(5.5^2)(1)}$$

$$= 819$$

Calculating the Dudley scoring index:

$$P = \frac{\text{number of teeth on pinion}}{d}$$

$$= \frac{24}{5.5} = 4.36$$

$$W = \frac{126,000(\text{hp})}{(\text{rpm})d}$$

$$= \frac{126,000(300)}{305(5.5)} = 22,533$$

$$\text{DSI} = \left(\frac{W}{F}\right)^{3/4} \frac{\text{rpm}^{1/2}}{P^{1/4}}$$

$$= \left(\frac{22,533}{10}\right)^{3/4} \left(\frac{305^{1/2}}{4.36^{1/4}}\right)$$

$$= 3950$$

The BSN value of 52,400 exceeds the maximum allowable value of 30,000 by a considerable amount. Notice that even a material change would only raise the allowable limit to 40,000. The TPI value of 819 also exceeds the maximum allowable value of 185 and is unacceptable. The DSI value of 3950 is considerably less than the 14,000 limit, so it is acceptable.

Even at the present rated conditions, the gears do not appear adequate. This is "raising the red flag" on an uprating. Several simplifying assumptions have been made in the BSN, TPI, and DSI equations, so a more detailed analysis would be appropriate, as would discussions with manufacturers on their calculations. It is quite possible that special conditions have been met on the steel quality and that the gears would pass the more detailed AGMA calculations.

Since the gears had failed several times in the past 15 years, a review of the failure history was made. This indicated that pitting had been a problem and failed gears with broken teeth were usually associated with a bearing failure. The outcome of this one-day analysis was that there exists the possibility that the gears cannot be uprated. A detailed analysis by the manufacturer was recommended, which confirmed the analysis.

Checking backlash: Verifying the backlash of a gear is important. Insufficient backlash can cause heat buildup due to interference of the teeth. This can result in a major gearbox failure. Backlash is checked by placing a dial indicator at the pitch line and rocking one of the gears through the clearance. The total amount of dial indicator movement is the backlash. In the absence of factory data, the following guidelines on what is acceptable backlash can be used. They are based on a gear center distance of 10 to 40 in.

For hobbed and shaped double helical:

$$\frac{\text{center distance (in.)}}{\text{backlash (in.)}} = 600 \text{ to } 800$$

For hardened and ground double helical:

$$\frac{\text{center distance (in.)}}{\text{backlash (in.)}} = 800 \text{ to } 1800$$

For hardened and ground single helical:

$$\frac{\text{center distance (in.)}}{\text{backlash (in.)}} = 600 \text{ to } 2200$$

The lower number is associated with the smaller center distances. For example, a single helical gear with a 10-in. center distance should have backlash of approximately $10/600 = 0.0167$ in.

Earlier in this section it was stated that the engineer should apply gear calculations on actual applications, both successful and not so successful. The following several installations show the BSN, TPI, and DSI values and how the gearbox performed. The methods are quite good for such comparisons. The important thing to notice is that gears that fail are usually considerably above the allowable limits for the cases tabulated, shown by an asterisk in Table 2.17. This illustrates the

TABLE 2.17   Documentation of Gear Failures

Gear Type	hp/rpm	BSN[a]	TPI[a]	DSI	Comments
1. Single helical	600/305	52,400	819*	3,950	Pitting, breakage history
2. Double helical	7,000/9,400	30,000	150	11,000	No failure history
3. Double helical	2,000/1,100	115,000*	500*	8,000	Pitting, breakage history
4. Double helical	800/1,200	8,000	90	4,000	No failure history
5. Herringbone	250/700	13,000	140	3,000	No failure history
6. Spur	15/1,800	39,000	210	1,600	No failure history
7. Single helical	450/6,000	15,000	110	6,000	No failure history
8. Spur	500/1,600	31,000	160	5.300	No failure history
9. Spur	50/30,000	33,000	550*	6,500	Wear failures
10. Spur	30/1,800	23,000	150	2,000	No failure history

[a]Asterisks indicate values above the allowable limits.

conservatism of the calculations and also shows that when the calculations indicate that a gear won't fail in a given mode, it usually doesn't. When the values are above the limiting values, gears usually fail. The engineer who accumulated these data would probably feel quite comfortable using the calculations as a screening tool or uprating comparison.

2.21   INTERFERENCE FITS

One method for assembling a shaft to a straight bore hub or for assembling a plug in a hole is to make the plug or shaft larger than the hole and force the shaft into the hub. Another method is to heat the hub so that the bore increases in diameter and then slide the shaft in. When the hub cools, the parts will be fastened. Calculation methods are required to answer the following questions when such interference fits are used:

- How much diametral interference fit is needed to hold the load and not overstress the hub?
- How much assembly force will it take to press the shaft in or out, or how much should the hub be heated?
- How much torque capacity can the fit transmit?

In Figure 2.39 we define some of the nomenclature used. The design equations for solid shafts in a hub are

$$\sigma_t = \frac{\frac{1}{2}E\delta}{D_s}\left[1+\left(\frac{D_s}{D_h}\right)^2\right]$$

$$p = \frac{\frac{1}{2}E\delta}{D_s}\left[1-\left(\frac{D_s}{D_h}\right)^2\right]$$

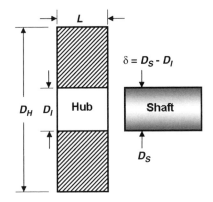

FIGURE 2.39   Hub and shaft nomenclature.

$$F = \pi f D_s L p$$

$$T = \tfrac{1}{2} F D_s$$

$$\delta = \alpha D_s \Delta T$$

$$i = (5.5 \times 10^{-8})(D_h^2 \times D_s)\left(\frac{\text{rpm}}{1000}\right)^2$$

where

σ_t = tangential hub stress (lb/in²)

$E = 30 \times 10^6$ lb/in² for steel

p = radial pressure between shaft and hub (lb/in²)

F = axial force (lb)

f = coefficient of friction: 0.15 normal metal to metal, 0.12 hydraulically fitted

T = resisting torque (in.-lb)

$\alpha = 6.6 \times 10^{-6}$ in./in./°F for steel

ΔT = temperature differential $(T_{\text{hub}} - T_{\text{shaft}})$ (°F)

i = loss of fit due to centrifugal effect in. on diameter

   One rule of thumb for interference fits is to use 0.001 in. of interference on the diameter for each inch of shaft diameter. From this rule, a 6-in.-diameter shaft would be assembled to a 5.994-in.-diameter straight hub. An approximation that can be made on complex or stepped hubs is to consider them as individual hubs. The torque holding ability is then the sum of the individual hubs.

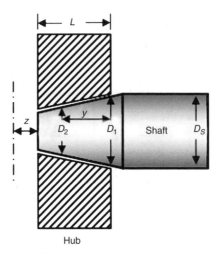

FIGURE 2.40   Tapered hub and shaft nomenclature.

Tapered shafts and bores are handled in much the same manner (Figure 2.40), with D_s now being the average diameter and δ the amount of interference fit caused by advancing the shaft on the hub by the amount z:

$$z = \frac{\delta}{\text{taper per inch}}$$

$$\text{taper per inch} = \frac{D_1 - D_2}{v}$$

After putting the hub on the shaft and marking it or placing a stop gauge on it, it would be pushed up the shaft by an amount z to achieve the interference fit desired. Taper is usually presented as taper per foot, with some popular values being $\frac{1}{2}$, $\frac{5}{8}$, $\frac{3}{4}$, $1\frac{1}{4}$, and $1\frac{1}{2}$. Here *taper* is defined as the larger diameter minus the smaller diameter over a 1-ft length. For example, if calculations indicated that a 0.002 interference fit on the diameter was required and a $1\frac{1}{2}$-in./ft taper was used, the design taper would be

$$v = \frac{1.5}{12} = 0.125$$

The axial travel z where the shaft would be marked after the hub was slid on push-tight would be

$$z = \frac{0.002}{0.125} = 0.016 \text{ in.}$$

2.21.1 Keyless Hydraulic Fitted Hubs

With tapered fits, a hub can be installed by heating it, then advancing it a prescribed amount. This increases the bore so that it can be positioned. As centrifugal compressors, gas and steam turbines and other high-performance machines increased in size, so did the required torque capacity of the hub fits, and heating of the hubs for installation caused problems.

Keyless hydraulically fit couplings provide a method for expanding the hub bore by pumping high-pressure oil between the tapered bore and tapered hub interface, usually in the range 35,000 lb/in^2. A specially designed coupling is used. When the hub is dilated sufficiently (i.e., expanded), it can be advanced along the taper using a special low-pressure hydraulic piston fixture at around 10,000 lb/in^2. This low-pressure fixture is removed and replaced with a nut. This is shown in Figure 2.41 only to illustrate; the actual components can differ considerably. The O-rings on the shaft hub are not necessary with proper design. Some of the advantages of eliminating these O-rings are:

- No grooves are required, which eliminates these stress risers.
- No grooves provide more gripping surface area, and less pressure is needed to expand the hub because of the increased area. The area is not limited to that between the O-rings.
- If desired, hubs can be installed by means of heat expansion and removed using hydraulics.

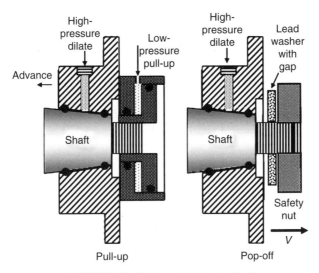

FIGURE 2.41 Fitting a hub hydraulically.

There are several reasons that keyless hydraulic fits are preferred over keyed fits:

- Greater torque transmission, because of the elimination of stress concentrations associated with keyways and fretting under the hub. Recall that stress concentrations and fretting of shafts and keyways are shown to be a major source of shaft failures.
- Minimal shaft damage from gouging, and no heating equipment is needed during hub installation and removal.
- Shorter installation and removal time.

There are several disadvantages to these types of couplings of which the user should be aware, but they are all manageable with special precautions and good procedures:

- Installation and removal can be hazardous and requires training and care, especially during removal.
- Design of the hubs is critical. The high dilation pressure can yield the hub and shaft ends if not designed and matched properly.
- Slippage is possible during motor acceleration and reaccelerations.

The importance of safety when removing these couplings cannot be overemphasized. When assembled, these couplings contain considerable potential energy, which is converted to kinetic energy when released. When not restrained, the hubs can fly off at velocity V. Unfortunately, there have been several very serious accidents to personnel due to this. *Never stay in the path that a coupling may take, even if there is a restraining nut, as threads can strip!* The case history in Section 2.21.3 illustrates the importance of safety.

Pumping up the pressure to remove a hub can be a hair-raising experience. The hub is released with a loud bang. To help cushion the impact, a lead washer can be used during removal, as shown in Figure 2.41. The small gap limits the velocity, and the washer deforms and absorbs the energy, resulting in a softer release.

Design Considerations   Reviewing the design of hydraulically fit couplings indicates that the industry seems to utilize the following practices:

- Diameter hub/diameter shaft ≈ 1.5.
- Length hub/diameter shaft ≈ 1.0.
- Fit ≈ 0.002 to 0.0025 in./in. shaft diameter, as more can grab the end.
- Friction coefficient ≈ 0.12.

2.21.2   Case History: Taper Fit Holding Ability

In this example, use of a motor to drive a centrifugal compressor was being evaluated. Taper fits are used on large-shaft, high-horsepower machines because they eliminate the need for high-stress-riser keyways, which had resulted in shaft failures. Also, they are easily installed and removed without heat and can transmit high horsepower and torque. The desire was to perform a quick calculation to verify the vendor's data at an early stage of the design. This example illustrates the use of most of the interference fit equations in this section.

We refer to Figure 2.40 on tapered fit nomenclature and insert the following data into the appropriate equations: $D_1 = 8$ in., $D_2 = 7.5$ in., shaft taper $= \frac{3}{4}$ in./ft $= 0.0625$, $L = 8.4$ in., and $D_h = 12$ in.

$$D_s = \frac{1}{2}(D_1 + D_2) = \frac{1}{2}(8.0 + 7.5) = 7.75 \text{ in.}$$

$$\delta = 7.75(0.002) = 0.016 \text{ in.}$$

$$z = \frac{0.016}{0.0625}$$

$$\Delta T = 303°F$$

$$\sigma = 44{,}000 \text{ lb/in}^2$$

A stress of 44,000 lb/in² indicates that yielding is not a problem.

$$p = 18{,}100 \text{ lb/in}^2$$

$$F = 555{,}300 \text{ lb} \qquad \text{axial holding capacity}$$

$$T = 2.15 \times 10^6 \text{ in.-lb} \qquad \text{torque holding capacity}$$

To determine if this is adequate, it must be compared with the worst-case torque that the interference fit must transmit. For the case of an induction motor, twice the nameplate rating during startup is not unusual. To be conservative, and to take into account the startup torsional vibrations and other unknowns, a factor of 3 will be used. This means that the torque capacity of the interference fit should have a design margin of at least 3, meaning that the holding capacity torque divided by the nameplate torque should be 3 or better to keep the coupling from slipping.

This motor has a nameplate rating of 7000 hp at 1750 rpm:

$$T = \frac{63{,}000(\text{hp})}{\text{rpm}}$$

$$= \frac{63{,}000(7000)}{1750} = 252{,}000 \text{ in.-lb}$$

The design margin is therefore

$$\frac{2{,}150{,}000}{252{,}000} = 8.5 \qquad \text{which is acceptable}$$

In the design equation 80% of the hub length L could have been used. This would accommodate a less than 100% contact of the shaft to the hub. The loss of fit i, due to the centrifugal effect, is negligible, due to the low speed.

2.21.3   Case History: Flying Hydraulically Fitted Hub

During the author's early career, he was unfortunate enough to witness a hydraulic hub being removed without a safety nut and without the proper precautions. With a loud bang the hub flew several feet; miraculously, no one was injured. That hasn't always been the case, however, and to demonstrate the energy stored in a hydraulically fitted or any tapered hub, an analytical model was developed. Although this model is rather crude, it does show that when suddenly released, the energy stored in the hub can propel the hub a good distance. The sudden change occurs when the hub is being removed and the fit is released, leaving just a lubricant film between the shaft taper and the hub taper.

Figure 2.42 depicts the analytical model used. For simplicity the hub is considered to be many radial elements, each forced out by an amount δ_R. The force acting on each element is the hub–shaft interface pressure. The displacement times this force represents potential energy and when all elements are summed up, it is the potential energy of the hub. The assumption is made that all of this energy is transformed into kinetic energy and the velocity of the hub determined. Trajectory calculations are then used to see how far the hub will "fly" with only the force of gravity acting on it.

Consider the case of the example hub in Section 2.21.2. The pressure at the interface is $p = 18,100$ lb/in^2, the length of the hub is $L = 8.4$ in., the hub inside diameter is $D_S = 7.75$ in., and the radial displacement $\delta_R = 0.008$ in. The weight of the coupling is $W = 120$ lb, and the hub centerline from the ground is $h = 48$ in.

The force on each element is the hub pressure times the area divided by the circumference of the hub inside diameter:

$$F_{element} = \frac{p\pi D_S L}{\pi D_S}$$
$$= pL$$

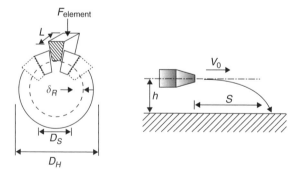

FIGURE 2.42   Hydraulic hub sudden release.

The potential energy per element is

$$PE = F_{element} \delta_R$$

This is the potential energy in the radial direction. The taper converts the components to the horizontal direction, but once the hub "pops" it doesn't follow the taper.

$$\delta_H = \delta_R$$

There is another sizable horizontal force acting on the hub, and that is the effect of the taper projected area and the hydraulic pressure on it. For a $\frac{3}{4}$-in./ft taper this is approximately

$$F_{horizontal} = \frac{\pi D_S (0.75 \times L/12)}{2} \times \text{hydraulic pressure}$$

For this example, this additional force will not be considered.

The total potential energy contained in the expanded hub is the sum of all the elements:

$$PE_{total} = pL\delta_R \pi D_S$$

Recall that the kinetic energy is

$$KE = \frac{1}{2} \frac{W}{g} V_0^2$$

Assume that all of the horizontal potential energy goes into kinetic energy. Equating the two equations and solving for the velocity yields

$$V_0 = \left(\frac{2pL\delta_R \pi D_S g}{W} \right)^{1/2} \quad \text{in./sec}$$

Notice in the equation that the smaller the hub weight W, the faster it will pop off when the friction holding force is released instantaneously due to the dilation pressure.

For the hub mentioned,

$$V_0 = \left[\frac{(2)18,100(8.4)(0.008)\pi (7.75)(386)}{120} \right]^{1/2}$$

$$= 436 \text{ in./sec or about 25 mph}$$

The distance S the 120-lb hub could travel from a height h until it hits the ground can be found from trajectory calculations. For an initial velocity of 436 in./sec and an initial height of 48 in., the distance the coupling will travel if not restrained is approximately

$$S = V\left(\frac{2h}{g}\right)^{1/2}$$

$$= 436\left[\frac{2(48)}{386}\right]^{1/2}$$

$$= 218 \text{ in. or } 18 \text{ ft}$$

This shows that you have to consider the zone in front of the coupling with respect, even with a safety nut attached. With the nut not fully engaged, threads have been known to strip because of the impact force of the coupling with the nut.

Although the method is very approximate and certainly should not be used to calculate safe distances, it does show the respect that should be given to tapered fit couplings of all types. Treat them as if they were a loaded gun and stay out of the line of fire.

2.22   STRENGTH OF WELDS

Welding is a common method for attaching parts, and usually cracks can be expected to develop at inadequate welds. Although there are all types of codes for the welding of steels that should be adhered to for critical structures, the following design stresses have been found useful for rough evaluations.

For steady loads:
  Butt welds: 13,000 lb/in^2 tension
              8000 lb/in^2 shear
  Fillet welds: 11,000 lb/in^2 all
  Under reverse loads: 5000 lb/in^2.
  Butt and fillet welds: 5000 lb/in^2

No stress concentration factors should be added to the stress equations that follow to use the design stresses. The load calculations can include a sudden load factor. In an earlier section, methods to calculate the loads (F) and moments (M) were shown. The following equations are useful for evaluating the stress (σ) and shear stress (σ_s) that occur in a weld [14]. The geometry and loading are shown in Figure 2.43. Several load cases are shown on each of the weld sketches.

For a tube or shaft fixed with a fillet weld size b:
  Stress in the weld due to moment M:

$$\sigma = \frac{5.66M}{\pi b D^2}$$

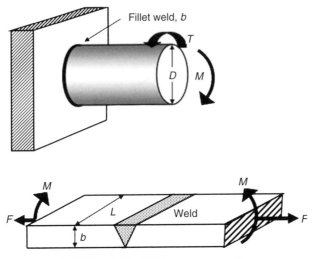

FIGURE 2.43 Weld stress geometries.

Stress in the weld due to torsional moment T:

$$\sigma_s = \frac{2.83T}{\pi b D^2}$$

For a plate with a full-penetration butt weld:
Stress in the weld due to load F:

$$\sigma = \frac{F}{bL}$$

Stress in the weld due to moment M:

$$\sigma = \frac{6M}{b^2 L}$$

The units used need to be consistent, such as in.-lb and in.

2.23 FATIGUE OF WELDS

Poor welds are a cause for fatigue-related failures (Figure 2.44). In Section 3.11, failures due to piping vibrations are shown. In vibrating conveyors, which are vibrating trays that transport product ranging from crumb rubber to hot castings, cracking of pans and welds are common, as shown by the case history in Section 10.6.2. These types of machines, unfortunately, are great fatigue test machines, as are welded

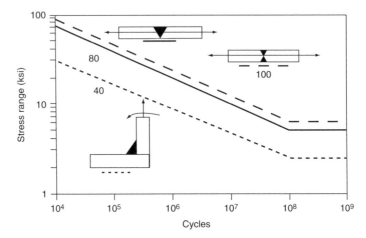

FIGURE 2.44   Weld classification and fatigue life.

vibrating piping. Only when the nominal stress-level range is kept below a stress range of 3 ksi can cracking of welds be greatly reduced. The API 579 [15] classifies welds and illustrates their fatigue strength.

Notice that the class 40 weld, which has an endurance limit of 2.2 ksi stress range at 10^8 cycles, has a built-in crack at the joint of the two plates. Most of the fatigue life goes into producing a crack, and since one already exists, it doesn't take much stress to propagate it. The class 100 weld, which is a full-penetration butt weld with a ground finish and is free of surface breaking defects, has an endurance limit of 6 ksi stress range at 10^8 cycles. This is an alternating cyclic stress of only ± 3 ksi, which indicates that even a very good weld does not like cyclic stress very much. This can be a real problem on pressure vessels that undergo cyclic pressuring up and blow-downs several times a minute. A misalignment of the welds can introduce bending stresses, which can result in fatigue cracking and subsequent leaks, or worse.

2.24   REPAIR OF MACHINERY

Machine design courses and books do not usually go into machinery clearances, fits, and repair techniques. For practicing engineers, at some time in their careers it will be very beneficial to have an understanding of what machine shops can do. From designing new machines to troubleshooting or repairing failures the following rules of thumb, many of them learned the hard way, should be useful. In general, they cover rotating machinery from 10 to 10,000 hp and speeds from 100 to 10,000 rpm.

2.24.1   Shafts

Shafts should be checked between centers and straightened if out by more than 0.001 in. Special straightening techniques are required for alloy shafts. All shaft shoulders for bearing fits should be checked with dial indicators for runout. Machine a shoulder if it is out by more than 0.0005 in. Undersized shaft fits can be repaired by chroming. Usually, this is limited to a 0.010-in. radial depth. An alternative method when only light compressive loads will be experienced is *metallizing*, which is sometimes referred to as *flame spraying*. A good general-purpose spray material is No. 2 stainless steel.

The ball or roller bearing inside diameter to shaft fit for machines such as general-purpose pumps is usually 0.0001 to 0.0007 in. interference fit on the diameter. Too much can preload the bearing and cause early failure. Too little will cause the inner race to move on the shaft and could result in fretting. If they are keyed and need to be removed without heat or a press, couplings should be metal to metal or 0.0005 in. on the diameter clearance. Such may be the case if they need to be removed on location.

2.24.2 Housings and Cases

Bearing outside-diameter bores that are oversize by more than 0.0015 in. can be repaired by boring the case or housing oversize and rebushing with the same material as the housing. A press fit of 0.002 in. interference on the diameter is usually adequate. Grind the bore to final size. The ball or roller bearing outside diameter to the housing should have 0.0005 to 0.001 in. of clearance on the diameter if the inner race is tight on the shaft.

2.24.3 Gearboxes

Gear center distances should be maintained to 0.002 in. or better on low-speed gear units. On low-speed gearboxes, the gear faces should be parallel to within 0.002 in./ft or better. All total indicator readings (TIRs) are usually held to 0.001 in. or less.

On split-line case distortion problems, shimming is sometimes successful, but often it is not, resulting in broken gear teeth or bearing distress. All that usually can be done with cast iron cases is to machine the split lines and rebore all fits. This lowers the shaft centerline, which must be considered on attached bearing plates and driven equipment. Such machining requires a high-quality machinist using a good boring machine. The setup is quite complex, and experience is a necessity.

2.24.4 Sleeve Bearing and Bushing Clearances

In many cases, 0.001 in. clearance on the diameter for each inch of shaft diameter seems to work well. For example, an 8-in. shaft would require an 8-in. + 0.008-in. bearing or bushing bore.

2.24.5 Alignments

Mechanical runout between two machines with a coupling that allows for some misalignment should not exceed 0.002 in. TIR. Use the reverse indicator method.

Loosening or tightening components such as piping should not change the alignment by more than 0.002 in. When doing alignments, don't forget that some equipment will change location with the difference in machine temperature at operating temperature and its cold condition. Most of the time, temperature is not a major concern and does not need to be considered. When the temperature difference is more than 200°F, the following rule of thumb is sometimes used on pumps and motors. The pump shaft is set 0.001 in. lower than the motor shaft for each 100°F above 200°F per inch distance between the pump base and the shaft centerline. For example, the calculation for a pump at 600°F with a 10-in. base-to-centerline distance would be

$$\frac{600-200}{100} \times (0.001) \times (10) = 0.040 \text{ in.}$$

To be at the correct alignment at an operating temperature of 600°F, the pump should be set 0.040 in. low or the motor set 0.040 in. high.

2.24.6 Acceptable Coupling Offset and Angular Misalignment

Figure 2.45, dealing with offset and angular misalignment, is for short couplings and is an average of many user experiences, one of which is referenced [16]. Couplings with spacer shafts will allow more angular misalignment. To minimize vibration and bearing loads, especially for machines operating at over 1800 rpm, some suggest that the offset and angular misalignment should be kept at 2 mils and 2 mils/in. or better, respectively.

2.24.7 Vibration Measurements

Velocity measurements in in./sec-peak are good indicators on the condition of many types of general-purpose machinery, such as pumps, low-speed gear units, motors, fans, mixers, small compressors, and internal combustion engines. Table 2.18 shows some indicators of the severity of vibration readings.

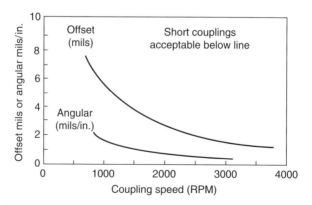

FIGURE 2.45   Allowable coupling offset and angular misalignment.

TABLE 2.18   Vibration Severity

Vibration Velocity (in./sec)	How Perceived
Up to 0.005	Very smooth
Up to 0.050	Good
Up to 0.100	Fair
Up to 0.200	Slightly rough
Up to 0.500	Rough
Over 0.500	Very rough

The most important thing to do when monitoring vibration is to have some idea of what the normal or baseline vibration of the machine is and look for deviations from this norm. The guidelines in the table are very crude. If some type of crushing equipment normally operates at 0.6 in./sec, this should be of little concern unless the vibration was gradually increasing. Similarly, if a motor is pretty smooth, as they usually are, and increases 0.1 to 0.2 in./sec, there is cause for concern.

2.25   INTERPRETING MECHANICAL FAILURES

Failed parts are trying to tell the observer something. Analyzing a failed part for the cause is an art from which the engineer benefits. Metallurgists are extremely well versed in this area, and when one is available, he or she should be utilized. Metallurgists can help determine the type of failure, what caused the failure, and what can be done to prevent such a failure, from a metallurgical point of view. When failures are due to corrosive effects on the material and alternative steels are sought, their input is especially helpful. The engineer often doesn't have access to this resource and needs an answer quickly. When this is the case, the following information should help.

2.25.1   Failures with Axial, Bending, and Torsional Loading

Knowing the mode of failure that has occurred allows a much better analysis to be performed because it can pinpoint the types of calculations that need to be made. For example, if a complex part shows evidence of an instantaneous failure, there is probably good reason not to waste time on fatigue calculations. The investigator should be concerned with what could cause such a sudden impact. Similarly, if fatigue due to a stress concentration is indicated, one should concentrate on reducing the stress riser. The following information and that of Figure 2.46 is based on areas that have helped the author analyze failures during his career. Much more detailed information is available in the literature [17,18].

One-Way Bending   A load on the end of a shaft that is not rotating would tend to see the following type of failure if the load is applied repeatedly in one direction

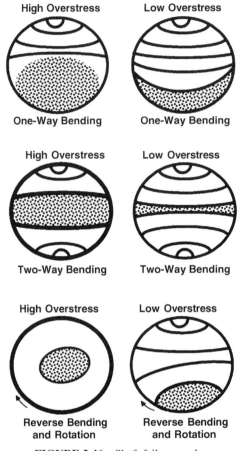

FIGURE 2.46   Shaft failure modes.

only, and released to its unloaded position. The two cases show the extremes of high and low overstress. The shaded zone is usually associated with a coarse high-overstress area and instantaneous fracture, and the curved lines are smooth, velvetlike fatigue areas. The fatigue crack started at the smallest semicircle and progressed slowly through these zones. Notice how on the high overstress fracture, most of the area has seen an instantaneous fracture. Due to the high stress, the crack progressed quickly once it was initiated. This coarse area is usually the zone that failed last, meaning that the last cycle caused the failure.

Two-Way Bending   This failure is similar to one-way bending, with the exception that the bending doesn't stop at the neutral position but also bends in the opposite direction. Because of the equal bending in both directions, the crack starts at about the same time on opposite sides and progresses toward the center. Unequal bending in each direction would cause one side to progress more quickly than the other: for

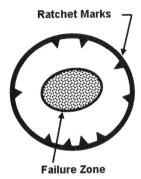

FIGURE 2.47   Shaft with ratchet marks.

example, if the load was higher in one direction than the other or there is a stress riser on one side. The crack would start sooner, so would propagate more on one side than the other. Again, no rotation of the part is present.

Reverse Bending and Rotational Loading   If the end of a shaft is forced up or down and rotated, as would occur on a rotating deflected shaft, the following pattern might be expected. The high-stress case would be expected if a sharp groove went around the shaft. Depending on the direction of the load, the coarse area could be centered or off-center. For an extruder shaft with high bending loads, the coarse zone was oval and near the outside diameter. Ratchet marks would probably also be noticeable on the high-overstress condition.

Ratchet marks   When a failure originates at more than one point, a phenomenon known as *ratchet marks* appears (Figure 2.47). This is usually associated with high stress. High stress in a snap ring groove with a sharp radius is an example.

Torsional failures   Brittle shaft failures usually fail in a characteristic 45° failure. The best way to visualize this is to twist a piece of blackboard chalk to failure and observe the failure mode. Usually, torsional failures are not so simple, due to combined loading. Unfortunately, several of these pure torsional failures have been observed on the main journals of crankshafts that had undergone severe torsional vibration problems. Torsional springs fail in this mode also. Ductile steels usually fail on a plane 90° to the centerline of the shaft.

2.25.2   Gear Teeth Failures

Gear Teeth Failures with Broken Teeth   When the end of a tooth is broken off, an analysis of the failure can be difficult, as the failure usually cascades. This means that the failed tooth works its way into the mesh and can strip a gear of its teeth. When a failure is caught early enough, failure across the root of the tooth is usually associated with a bending type of failure. When a clean break occurs completely

FIGURE 2.48   Pinion gear with broken teeth.

across the root, this is an indication that the tooth was not strong enough or was over-loaded. A calculation of the bending strength number would be advantageous at this point, along with a review of possible overload conditions or history. Since failures can also develop at defects such as pits, the pitting index should also be checked, especially if the tooth broke off near the tooth line of contact.

When only the corner of the tooth is broken, as shown in Figures 2.48 and 2.37, the possibility of gear misalignment is present. Such misalignment causes only the end of the tooth to carry the load, thus the corner breakage due to the small beam strength. Essentially, instead of having the full tooth face width in the bending strength equations, only 10% or less would be active.

Gear Teeth Failures with Pitting and Spalling Along the Pitch Diameter   *Pitting* is generally thought of as small pits or holes along the pitch diameter of a gear. The closeness and density of the pits are indications of how the load is distributed. *Spalling* is an advanced form of pitting in which large pieces of material appear to have been removed from the surface. Both types of failures are fatigue failures and are usually associated with high loading. Figure 2.49 illustrates pitting and spalling distress due to a failed bearing.

Very early in the author's engineering career he was given the assignment of witnessing the startup of a new gearbox of a double-helix type. After an extended run the cover was removed and the gear pattern observed. There was some fairly uniform pitting along the pitch line, certainly not severe. The gear manufacturer's representative said not to worry, this was *arrestive pitting*, which equalized the loading and would not progress further. A month later we looked at it again and were now into what he termed *destructive pitting*, which he felt would keep progressing.

The moral of the story is to consider any pitting as a problem and to try to identify a cause. Bring in outside gear experts if you are in doubt. Deflection under load, wrong gear material heat treating, machining errors, bearing problems, and improper startup can all be contributors to pitting and spalling.

FIGURE 2.49   Gear with pitting and spalling distress.

2.25.3   Spring Failures

Spring failures are usually torsional-type failures that are due to the loading. Springs
are bars in torsion. Figure 2.50 is for the attachment end of a large spring for a vibratory drier. It failed because of incorrect heat treatment from an inexperienced supplier. Since most springs are of hardened steel, the normal *beach marks* or striations
seen in ductile failures are not always present, except for steels with a hardness

FIGURE 2.50   Spring failure due to wrong heat treatment.

below Rc 45. This can cause the inexperienced to come to incorrect conclusions. Most failures are fatigue related, and many are due to surface defects or corrosion on a highly stressed spring. Other causes are defective heat treatment from the supplier, or just a poorly designed spring that is overstressed during operation.

Normally in torsion, failures will occur 45° to the spring axis. If the spring was in bending, it would be at 90° to the spring axis. However, this is quite dependent on the hardness or surface treatment. Spring failures with significant consequences should be analyzed by a laboratory and the failure characterized as to the material type, hardness profile, and failure cause.

2.25.4   Bolt Failures

Many bolt failures that occur are from overloading or from an immediate failure caused by an impact load (Figure 2.51). Cyclic loading where the bolt fails in fatigue usually at the first thread, or from a corrosion mechanism such as stress corrosion cracking (SCC), is also quite common. What the failure zone will look like depends on whether the bolt is made of a ductile or a brittle material. For this particular discussion of bolting, consider a material of Rc 15 to be ductile and Rc 60 to be brittle. A tension failure of a ductile bolt could be expected to have a section of reduced area, or be *necked down*. A brittle bolt would have a failure with no visible area reduction, probably through the first thread. The failures can be analyzed for fatigue in the manner mentioned earlier in this section.

On the left of Figure 2.51 is a $\frac{7}{8}$-in.-diameter bolt of 4140 material Rc 45. It was a short cap bolt that was holding a cycling hydraulic cylinder. The bolt loosened, a crack started at the 12 o'clock position, and the bolt failed under the bolt head due to a sharp radius. For the inexperienced, the wrong conclusions might be reached as to the cause of the failure.

On the right in Figure 2.51 is shown the bolt used to secure blades to an agitator (discussed in Section 10.3.1 as a case history). The blades were struck by product chunks and the short bolts were loosened. Beach marks and initiation points are

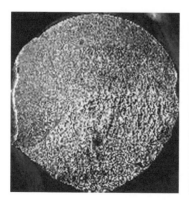

FIGURE 2.51   Bolt failures due to fatigue.

evident at the 5 and 7 o'clock positions, as is the fact that there were two initiation points, so impact and fatigue were at work on this failure.

Failures from a corrosion mechanism such as SCC usually require a metallurgical examination to identify the problem. Several factors are necessary to produce such a failure. A tensile stress, which could be a service-induced or a residual stress, a susceptible alloy, and a specific environment are all necessary.

It is difficult to identify which materials will be susceptible to SCC. In the petrochemical industry, when SCC is a problem, the solution usually centers on using a specialized bolt material such as Inconel Alloy 718 or its equivalent, which has high-strength characteristics and corrosion resistance. Although the cost of such alloys may be higher than that of other steels, historically it has performed well. Other obvious fixes would be to lower the tensile stresses by design changes such as rolled threads instead of cut threads, or to eliminate the chloride environment.

When a bolt has failed under the bolt head of a high-alloy bolt, causes such as bending due to nonparallel surfaces have been identified. In one instance, a bolt was failing in fatigue and a high-alloy bolt was utilized and torqued to a much higher value. The bolt continued to fail because of the high bending stress induced in the radius connecting the head to the bolt body. This was due to the high preload, with a point contact under the bolt head which caused the bending moment. The bending moment magnitude was approximately the preload force in the bolt times one-half the bolt head diameter. Using a spherical washer under the bolt head solved the problem by removing the bending effect.

2.25.5 Bearing Failures

Major manufacturers of bearings are masters at defining what the cause of a bearing failure was even when the bearing looks as if it has been totally destroyed. By looking at the ball tracking on the race it can be determined if the bearing was cocked, had excessive thrust load, had excessive preload, was distorted, or suffered from one or more of a multitude of other possible causes [19]. For critical bearing failures, consultation with the bearing manufacturer's engineering department is recommended.

Since our concern is with life and loads, we look at the following type of failure, which occurs often and is similar to the pitting and spalling failures discussed in Section 2.20. When it occurs, a load–life calculation would be wise. Figure 2.52 represents a roller element bearing that pitted and spalled due to an excessively high thrust load due to a tight fit in the housing and on the shaft. It is analyzed as a case history in Section 10.3.13.

Figure 2.53 is a bearing that was affected during harsh startup conditions. It is analyzed in Section 10.3.9. The majority of bearing problems on existing equipment are usually related to a lack of lubrication. Failures that occur early in a machine's life after a repair are usually due to a lubrication failure, assembly damage, contamination, or using the wrong fit. The point is to look for the obvious first and not be too concerned about load calculations, unless the obvious are not a factor.

As an illustration of the effect of lubrication, consider Figure 2.54, which is based on one plant's actual data and shows bearing life as a function of the rolling element

FIGURE 2.52 Bearing failure due to excessive load.

FIGURE 2.53 Bearing failure due to impact.

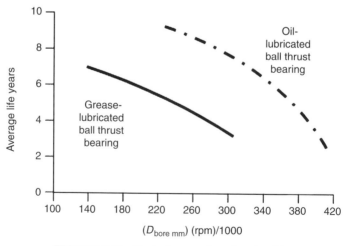

FIGURE 2.54 Vertical motor thrust bearing life.

velocity and lubrication method. Knowing the bore, speed, and lubrication method provides information on the average bearing life. The graph represents 160 different motor bearing failures. Approximately 85% of the failures were defined as a lack of lubrication. This is understandable since higher speeds and larger-diameter bearings will require shorter lubrication intervals, some as short as monthly. The need for the addition of lubrication is obvious with an oil bath system since an oil-level gauge is usually installed. The large oil sump does not require frequent replenishing. Grease-lubricated bearings require a systematic greasing schedule to be maintained. This is usually difficult to keep implemented, as it requires the work of someone other than the operators. The result is that greased bearings fail more often than oil bath bearings.

2.25.6 Reading a Bearing

When a rolling element bearing rotates under load, the contacting surfaces generally become dull in appearance, which is normal and doesn't indicate abnormal wear. The raceway's dull surface forms a loading pattern, which varies in appearance according to rotational and loading conditions. By examining this pattern closely, it is possible to determine if the bearing is operating normally or abnormally. Major bearing manufacturers have many of these types of charts for use in troubleshooting bearing problems.

Although the method appears quite simple, in reality it is not. Most bearings are damaged to such an extent that the patterns are not easily interpreted. The charts are of most use if you are lucky enough to observe the bearing in the early stages of failure. The following represent some of the types of failures that occur more often in ball bearings.

Under normal radial loading with either a rotating inner or outer ring, as shown in Figure 2.55, the pattern is centered and is about the same width and heaviness around the race. Under normal combined axial and radial loading, the patterns are offset as shown in Figure 2.56. From the load line and stars, which are the load reaction points, it is evident how the pattern is generated. For the case where the inner race is rotating, the outer race is fixed, and the inner race is misaligned, the pattern

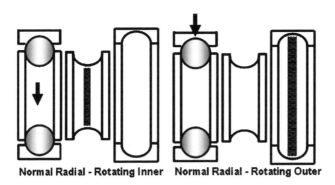

Normal Radial - Rotating Inner   Normal Radial - Rotating Outer

FIGURE 2.55   Bearing with normal radial load.

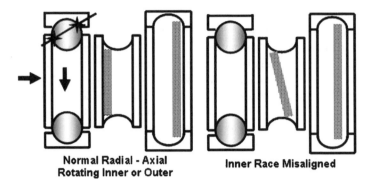

**Normal Radial - Axial
Rotating Inner or Outer**   **Inner Race Misaligned**

FIGURE 2.56   Bearing with normal axial and radial load and misaligned.

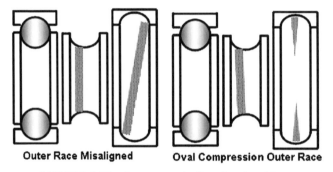

Outer Race Misaligned   **Oval Compression Outer Race**

FIGURE 2.57   Outer race misaligned and oval bore.

becomes diagonally displaced as shown. A failed coupling might present this appearance if observed early in the failure. When the outer race is misaligned as in Figure 2.57, it is the outer race that becomes diagonally displaced.

Bearings that are fitted to oval bores will be squeezed in where the bore is minimal. This will be observed as a wide pattern where this pinching has occurred. The top and bottom of the oval compression outer race bearing were pinched in the most in the sketch, and the triangular pattern is as shown in Figure 2.57.

This discussion covers only a couple of cases that have occurred several times that the author has found useful. The heaviness of the pattern or initial pitting or spalling will also be useful in revealing where the heavy loads have occurred. Many of the other patterns are best left to bearing experts to examine and diagnose. Most of the failures the author has been involved in were beyond being in the initial failure region. Many were not even recognizable as bearings. This is usually the time when a good failure investigation, including analytical analysis of the loads, is required.

2.25.7   Large Gearbox Keyway and Shaft Failures

The keyway and shaft failures observed usually have considerable fretting in the keyway and on the shaft surface, as shown in Figure 2.58. This is often due to a poor fit of the key in the keyway after a repair or in the initial design. Fretting is caused by a minute sliding motion of the key during operation, which can cause a lowering of the surface fatigue strength of the shaft. The key should be a snug fit in the keyway so that such motion does not occur. The greatly exaggerated loose key shown in Figure 2.59 will load up on one side of the keyway. Eventually, a fatigue crack will start and propagate though the shaft. A keyway with sharp corners (i.e., no radius) is also likely to start a crack at this high stress point. When caught early, sometimes only a section of the shaft may break out, as shown.

FIGURE 2.58   Fretting and failure of a shaft keyway.

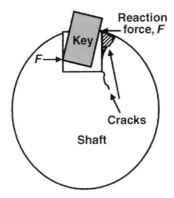

FIGURE 2.59   Loose key reactions and breakout.

2.26 CASE HISTORY: SIZING A BUSHING RUNNING CLEARANCE

This example involves the use of a very simple thermal expansion equation. The interest is in the extreme temperature variations that the shaft, bushing, and housing undergo, and the objective is to size a bushing clearance so that it is tight enough to limit the shaft's motion, but loose enough not to seize on the shaft, due to lack of clearance. The clearance loss occurs due to temperature changes from $-150°F$ to $+170°F$. The bearing material is a graphite–babbitt material, and the installation was in a reactor with methyl chloride on the cold cycle and wash oil on the hot cycle.

The accuracy of any analysis of this type is based on knowing the temperature of the parts. For this analysis the assumption was made that the components are at the product temperature. The only assurance that this is an adequate assumption is that the calculated clearance did not show any indication of seizing the shaft after considerable service.

Some interesting points of this analysis are that:

1. If you did not consider temperature and sized the bearing clearance at 0.005 in. on the diameter when assembled, it could fail by seizing the shaft to the bushing.
2. The cold condition is the controlling condition to design to since it results in the largest clearance loss (0.0053 in.).
3. Although not shown in this example, the loss or increase in the press fit of the bushing under operating conditions can be calculated using similar techniques.

The solution is based on the model shown in Figure 2.60.

$$\text{Diametrical clearance} = D_{\text{bid}} - D_S$$

$$\text{Controlling equation } \delta = \alpha D \, \Delta T$$

$$\alpha = \text{coefficient of expansion (in./in.-°F)}$$

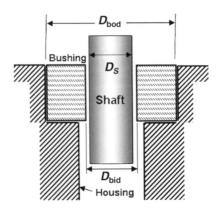

FIGURE 2.60 Thermal expansion model.

$\quad D =$ diameter (in.)

$\quad \Delta T =$ temperature difference of parts from startup condition (°F)

$\quad \delta =$ shrinkage ($-$) or expansion ($+$) on diameter (in.)

Under cold conditions:

$$\Delta T_C = 70° - (-150°F) = 220°F$$

Housing and bushing shrink:

1. Bushing installed in housing 0.003 in. interference fit. Assume that D_{bid} decreased this much.
2. Housing shrinks compressing bushing:

$$\delta_{bod} = \alpha_H D_{bod}\, \Delta T$$
$$= 7 \times 10^{-6}(2.5)(220°F)$$
$$= 0.0039 \text{ in. shrink}$$

3. Bushing ID shrinks ($D_{bid} = D_S$):

$$\delta_{bid} = \alpha_B D_S\, \Delta T$$
$$= 2 \times 10^{-6}(1.4)(220°F)$$
$$= 0.0006 \text{ in.}$$

Shaft shrinks:

$$\delta_S = \alpha_S D_S \Delta T$$
$$= 7 \times 10^{-6}(1.4)(220)$$
$$= 0.0022 \text{ in.}$$

Running clearance at temperature:

$$\delta_{running} = D_{bid} - D_S$$

δ_{bid} inward housing shrink (lose) $= -0.0039$

δ_{bid} due to press in fit (lose) $= -0.0030$

δ_{bid} due to bushing ID thermal (lose) $= -0.0006$

Shaft shrinkage (gain) $= +0.0022$

Total loss $= 0.0053$

For a desired running clearance of 0.005 in., the initial clearance needs to be 0.0053 + 0.005 = 0.010 in. on the diameter between the shaft and bushing when cold.

Under hot conditions:

$$\Delta T_H = 70 - 170 = 100°F$$

Housing and bushing expand:

1. D_{bid} decreased 0.003; same as cold.
2. Housing expands.

$$\delta_{bod} = 7 \times 10^{-6}(2.5)(100)$$
$$= 0.0018 \text{ in.}$$

3. Bushing ID expands:

$$\delta_{bid} = 2 \times 10^{-6}(1.4)(100)$$
$$= 0.0003 \text{ in.}$$

Shaft expands:

$$\delta_S = 7 \times 10^{-6}(1.4)(100)$$
$$= 0.001 \text{ in.}$$

Clearance required:

$$\delta_{bid} \text{ housing expansion (gain)} = +0.0018$$
$$\delta_{bid} \text{ due to press fit (lose)} = -0.003$$
$$\delta_{bid} \text{ due to bushing ID (gain)} = +0.0003$$
$$\text{Shaft expansion (lose)} = -0.001$$
$$\text{Total loss} = -0.0019$$

Under hot conditions, if the required running clearance is again 0.005 in., the initial clearance needs to be

$$0.0019 + 0.005 = 0.007 \text{ in.}$$

The cold condition controls, so the clearance on the diameter should be at least 0.010 in. The bushing inside diameter should be machined to

$$D_{bid} = D_S + 0.010 \text{ in.}$$

2.27 CASE HISTORY: GALLING OF A SHAFT IN A BUSHING

A good idea when troubleshooting failures is to question everything you are told. This particular case involves the failure of a large vertical process pump. It had an internal bronze bushing that was used as a steady bearing for a 4340 steel shaft, as shown in Figure 2.61. The shop report indicated that the shaft and bushing were badly galled. The consensus was that the shaft was striking the bushing due to operational upsets that caused the galling to occur. The bushing/shaft clearance was tightened from 0.005 in. to 0.003 in. on the diameter to reduce the effect of impact. After startup, the same type of failure occurred within two weeks. At this point the credibility of the repair shop's reasoning was questioned and the troubleshooting began.

Galling is a sort of welding of one part to another. For it to occur, there usually has to be sufficient force along with materials that are susceptible to galling. Lack of lubrication increases the tendency to gall. Table 2.19 shows some material combinations to avoid. The letter L indicates that galling is likely to occur, and NL, that it is not likely to occur. For example, a nonlubricated 316 stainless steel stud should not be used in a threaded 316 stainless steel plate or nut and expect them to unthread easily.

However, when we look at bronze against a hard steel, which 4340 is, galling is also shown to be unlikely. A second verification of this was that the bronze bushing didn't really appear to be beaten about as much as it looks burned. Since it probably is not galling due to loads in contact, the wisdom of tightening up the shaft to the bushing clearance was also questioned. A major clue occurred when an operator mentioned that the process has had severe temperature excursions.

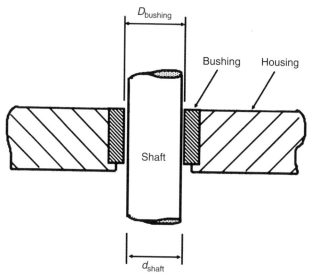

FIGURE 2.61 Shaft in bushing galling.

TABLE 2.19 Galling Tendencies of Metals

	Cast Iron	300 Series SS All	400 Series SS Soft	400 Series SS Hard	SAE 1000 to 6000 Soft	SAE 1000 to 6000 Hard	Bronze	Nitride	Stellite	Chromium Plate
Cast Iron	NL									
300 Series SS	NL	L								
400 Series SS Soft	NL	L	L							
400 Series SS Hard	NL	L	L	NL						
SAE 1000 to 6000 Soft	NL	NL	L	NL	L					
SAE 1000 to 6000 Hard	NL	L	L	NL	NL	NL				
Bronze	NL	NL	NL	NL	NL	NL	NL			
Nitride	NL	NL	NL	NL	NL	NL	NL	NL		
Stellite	NL	NL	NL	NL	NL	NL	NL	NL	NL	
Chromium Plate	NL	NL	NL	NL	NL	NL	NL	NL	NL	NL

The process temperature excursions were recorded as going to 170°F, which can also be assumed to be the shaft temperature. The bushing fit was metal to metal with the housing, and the housing was massive and stays at 100°F even during the upset. It simply cannot react to the temperature change as quickly as the shaft can react. One can speculate that since the shaft is hotter than the housing, it increased in diameter enough to bind in the bushing during the temperature upset. This would explain the burned appearance of the bushing due to the high friction causing heat. The following calculation was used to validate this possibility.

The shaft diameter expands

$$\delta = \alpha_{steel} d_{shaft}\, \Delta T \qquad \text{in.}$$

For this case, $\alpha_{steel} = 6.6 \times 10^{-6}$ in./in.-°F., $d_{shaft} = 6$ in., $\Delta T = 70°F$, and thus

$$\delta = 0.0028 \text{ in.}$$

This is much too close to the assembly clearance of 0.003 in., and it is quite likely that all clearance was lost between the shaft and the bushing. Tightening up the clearance from the original 0.005 in. was not the solution, nor was galling the problem. The situation was corrected by returning to the 0.005-in. clearance and maintaining better control on the temperature to prevent such large deviations.

2.28 CASE HISTORY: REMAINING FATIGUE LIFE WITH CYCLIC STRESSES

Earlier, when fatigue was discussed along with the modified Goodman diagram, the alternating stress magnitude was assumed constant throughout the life of the component. This is not always the case, as might be expected in the fatigue of aircraft wings due to wind gust loading, or in automobile suspension parts due to road profiles. In this example, the alternating stress magnitude changes with time, and the fatigue life exhausted was required. Most involved with machinery have heard a comment such as: "Yes, it was overloaded, but the machine is back to normal now, so no harm was done." For machinery this is not necessarily so, as this case history shows.

Vibrating conveyors are used to transport difficult-to-handle products, and this one had failed after only three years of service. Machine designers keep the operating stresses very low in vibrating conveyors because if overstressed, they are excellent weld fatigue test machines. This particular conveyor operates at 500 strokes/min (30,000 cycles/hr). With a three-year run length before a failure occurred, design flaws were quite unlikely since almost 1 billion cycles were on the machine. If it was going to fail in fatigue, it would have done so long ago. What was the cause? A review of the conveyor history revealed some interesting data. Twice a diverter gate was not tightened securely and was allowed to hit the stops. This impact increased the cyclic stresses in certain welds and was accompanied by considerable

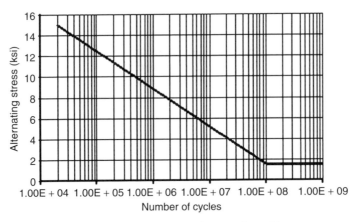

FIGURE 2.62   Fatigue life curve for weld.

noise, which was obvious to the operators. When the gate was secured, the stresses dropped back to normal and no damage was indicated.

Most of this conveyor's three-year history was at the design alternating stress of 1500 lb/in². During the second year, $\frac{1}{2}$ hour (15,000 cycles) of impact was endured at 15,000 lb/in², and during the third year it occurred again for $\frac{1}{2}$ hour at 12,000 lb/in². The stresses were determined from strain gauge tests at a later date, when the conditions were simulated.

The next piece of information needed to determine the cause is the fatigue curve of the material in its welded condition. This particular curve (Figure 2.62) is from actual operating experience on conveyors that had cracked in the welds. A method to determine about how much of the fatigue life is used up by these brief spurts of overstress, called the *Palmgren–Miner linear damage rule* [20], is represented by the equation

$$\Sigma \frac{n_i}{N_i} = \frac{n_1}{N_1} + \frac{n_2}{N_2} + \cdots = 1$$

N_1 and N_2 are from the fatigue curve and represent the life at that stress level, and n_1 and n_2 are actual cycles at that stress level:

$$\frac{15,000}{150,000} + \frac{15,000}{20,000} + \frac{n_3}{N_3} = 1$$

$$0.10 + 0.75 + \frac{n_3}{N_3} = 1$$

As shown on Figure 2.62, a stress of 1,500 lb/in² or less in the welds should not result in a fatigue failure. However, from the equation, 10% of the life was used by overstressing to 12,000 lb/in², and a whopping 75% additional life was used up by overstressing to 15,000 lb/in². The remaining n_3/N_3 or 15% was probably from some unrecorded incident.

The important lesson is that the conveyor didn't "forget" harsh treatment and the expected life was greatly shortened. The higher the overload, the shorter the life. Explaining that a machine "remembers" can sometimes expedite response times to abnormal operations by the machine operator. However, telling the operator "it remembered" can also generate some very strange looks from those who may question your sanity.

2.29 PROCEDURE FOR EVALUATING GASKETED JOINTS

In the area of pressure vessel design, it is important to keep gasketed joints from leaking. Leaking joints can be an economic as well as a safety issue. In this section we present a quick method for evaluating the tightness of the joint and what to look for if unacceptable leaks occur.

For a gasket to be effective, it must seal the joint between surfaces. If the surfaces were perfectly flat and parallel with a mirrorlike, defect-free finish, theoretically, a gasket would not be required. A bolt load to keep the two surfaces in contact would be adequate. This is not the case, and gaskets are required so that one surface can conform to the other. Gaskets such as spiral-wound units provide some resiliency in case the bolting or surfaces relax. Many gasket materials are more like an inert filling material since their elastic properties are doubtful when they are highly compressed. This means that they do not rebound much as a joint relaxes and is one reason that retightening is usually a successful procedure for preventing leaks. The guiding principle of a tight joint is to design it so that the bolting force supplied is sufficient to seat the gasket and prevent the operating pressure from opening it. This was observed in Section 2.17.7, where a power head was leaking.

One design approach that is simple to use is an abbreviated version of the ASME Boiler and Pressure Vessel Code [21], a procedure for evaluating gasketed joints. The method presented here is only for quick evaluations; the latest code should be used for design purposes. The code does not recommend a bolt load, so in this simplified analysis a bolt stress of 50,000 lb/in² is used. This value has had considerable success in industry for keeping joints tight. The limiting factor can be distortion of flanges by too high a bolt load. The detailed calculation procedure should be used in design work, as 50,000 lb/in² could result in distortion of certain types of flanges.

The force required to overcome the operating pressure is

$$W_{m1} = \frac{\pi}{4} G^2 p + (2b\pi Gm)p$$

In the equation above, p is the process pressure in lb/in², and m can be thought of as a test factor found necessary to keep the joint tight. Notice that the first term is simply the pressure load applied to the joint. The term in parentheses is the surface area in contact with the gasket times m.

The force needed to seat the gasket is

$$W_{m2} = \pi b G y$$

In the equation above, b is the effective gasket seating width (in.) which is about half the actual width for raised flanges, G is approximately the midpoint of the gasket diameter, and y is the stress (lb/in^2) required to seat the gasket. So the force W_{m2} is the gasket area times the stress to yield it.

The total bolt load available is

$$W_{total} = 39{,}250D^2(\text{number of bolts}) \qquad \text{lb}$$

This load is based on producing 50,000 lb/in^2 nominal stress in each bolt of diameter D (in.). The torque necessary to produce this stress on lubricated threads (not Teflon) is approximately

$$\text{torque} = 393D^3 \qquad \text{ft-lb}$$

The joint is considered sealed when W_{m1} and W_{m2} are each less than W_{total}. Typical gasket factors from the ASME code and vendor catalogs [22] are shown in Table 2.20.

Nominal gasket stresses (i.e., the pressure on a gasket) are sometimes helpful in troubleshooting. Too high a stress and a gasket could crush, and too low a stress and it might not seal. When graphite gaskets have too high a stress, the surface has the appearance of a corrosion type of mechanism. Overstressing jacketed gaskets can rupture the jacket.

Nominal gasket stress at assembly: When a stress other than 50,000 lb/in^2 is used, multiply σ_{gasket} by

$$\frac{\text{bolt stress (lb/in}^2)}{50{,}000}$$

TABLE 2.20 Gasket m and y Factors

Type of Gasket	m Factor	y Factor
Elastomer	1.25	400
$\frac{1}{16}$-in. asbestos sheet	2.75	3,700
Spiral-wound, asbestos filled	3.00	10,000
Stainless jacketed, asbestos filled	3.75	9,000
Solid flat copper	4.75	13,000
Graphite laminate		
$\frac{1}{16}$ in.	2.60	1,700
$\frac{1}{8}$ in.	6.00	3,000
PTFE varies with binder	3.00	3,000

For the 50,000-lb/in² case,

$$\sigma_{\text{gasket}} = \frac{W_{\text{total}}}{(\pi/4)(D^2_{\text{ODgasket}} - D^2_{\text{IDgasket}})}$$

In Section 10.6.4 a value of four times the internal pressure is mentioned as a gasket stress value the author has used that usually doesn't result in flange leaks. This is used for quick troubleshooting checks and is not intended for design purposes.

Gasket Applications   Figure 2.63 summarizes some typical gasket types, and Table 2.21, typical applications.

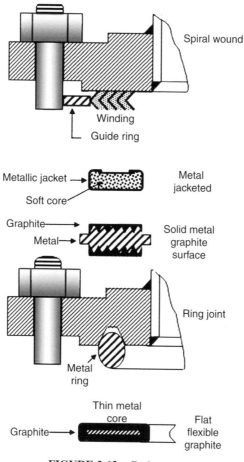

FIGURE 2.63   Gasket types.

TABLE 2.21 Typical Gasket Applications

Service	Flange (lb)	Temperature (°F)	Type
Many hydrocarbon services	600	−20 to 700	Spiral wound-flexible graphite filled with inner ring
Nitrogen, air, water	150	−20 to 225	White neoprene or PTFE
Many hydrocarbon services	150	−150 to 400	Corrugated metal graphite, covered
High-pressure service	150	To 900	Soft-iron oval ring

2.30 GASKETS IN HIGH-TEMPERATURE SERVICE

When temperatures are outside the range of conventional gasket materials, ring joints, metal–silver overlays, spiral-wound or metal gaskets with micalike fill, or welded gaskets are used. Figure 2.64 shows the effect of using nonasbestos graphite gaskets in high-temperature services. Using specially filled or enveloped graphite gaskets can overcome some of these limitations.

Load reduction can cause gasket leaks during thermal cycles, such as startup and shutdowns or temperature swings in the process. The gasket does not retain enough resiliency to spring back and maintain the seal. Other areas of concern that need to be considered in high-temperature service are:

- Flange distortion due to warping
- Bolt–flange thermal compatibility

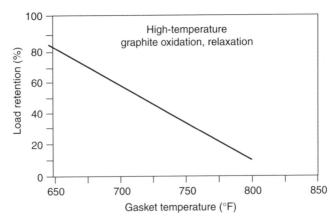

FIGURE 2.64 Graphite gaskets in high-temperature service.

- Flange insulation and rain effect
- Bolt yielding due to overload

In summary, high-temperature joints need to be well engineered.

2.31 O-RING EVALUATION

O-ring sealing components are use extensively on mechanical, oil and gas seals, hydraulic cylinders, pumps, compressors, and just about all machines that need to seal fluids or gases. Although they are relatively troublefree, they can result in leakage failures, such as in pump mechanical seals. In this section we deal briefly with the types of materials used most commonly in the manufacture of O-rings and their compatability with fluids.

Nitrile rubber, BUNA-N Service to 225°F (107°C) is used extensively in the oil field and petroleum industry because of its general resistance to petroleum-based liquid hydrocarbons. It is not well suited for acetic acids, chlorine, some Freons, hydrogen sulfide, steam, benzene, and methyl ethyl ketone (MEK), to name a few.

EPDM (ethylene propylene rubber) Usually limited to service below 400°F (204°C) and not well suited for gasolines, greases, oils, many Freons, LPG, butanes, and others.

Viton (fluoroelastomer, Dupont) Usually limited to service below 400°F (204°C) and not well suited for acetic acids, acetones, anhydrous ammonia, Freons, MEK, steam, and others.

Kalrez (perfluoroelastomer, Dupont) Usually limited to service below 500°F (260°C) and not well suited for Freons or fluorines.

These are only a few of the materials used; to be sure of an O-ring's compatibility, check with manufacturers or suppliers. Table 2.22 explains some failure modes of O-rings. When an O-ring cracks, hardening or oxidative cracking caused by excessive temperature may have been the cause. A materials laboratory familiar with nonmetallics will be able to characterize the failure.

The author had the experience of helping replace a horizontal cylinder on an aircraft engine. There was a very slight oil leak at the base of the cylinder to the crankcase, which was sealed with an O-ring. The leak was caused by a scratch on the surface no more than 0.005 in. wide and 0.005 in. deep. The O-ring just bridged the gap and would not deform into it to provide a seal. Polishing out the scratch solved the problem. It doesn't take much of a surface imperfection to cause a leak.

Sometimes there is no visible sign of failure when examining an O-ring. When this occurs, eccentric components may be the cause, or no compression of the O-ring could be the problem, due to tolerance stack-up or improper sizing.

TABLE 2.22   Causes of O-Ring Failures

Failed Shape	Cause
	Compression set: • Both sides of cross section flat Typical cause: • Excessive squeeze from overtightening gland • Excessive temperature caused swelling, losing elasticity • Volume swell due to system fluid
	Extrusion/nibbling: • Chips taken out of edges normally on the low-pressure side Typical cause: • Degradation caused by system fluid • O-ring too large for groove • Excessive clearances or pressures • Sharp edge on gland • Too soft an elastomer
	Installation damage: • Usually seen as flats, short cuts, notches, skinning, or peeling on the surface Typical causes: • Assembly carelessness, pinched, twisted, no lubrication • Insufficient chamfer or sharp edge • Too large or too small an O-ring

2.32   CASE HISTORY: GASKET THAT WON'T PASS A HYDROTEST

To show the use of the method in Section 2.29, consider a large vessel that is about to be delivered for a new project. The schedule is tight and the shop can't get the vessel to seal during a hydrotest. Pertinent data are:

Spiral gasket $\quad D_{OD} = 86.625 \quad D_{ID} = 84.625$
$m = 3 \qquad\qquad y = 10{,}000 \qquad D_{bolt} = 1.375$
88 bolts $\qquad\quad p = 600 \text{ lb/in}^2 \quad b = 0.5 \text{ in.}$

$$G = \frac{86.625 + 84.625}{2} = 85.625 \text{ in.}$$

$$W_{ml} = \frac{\pi}{4} G^2 p + (2b\pi Gm)\, p$$

$$= \frac{\pi}{4}\, 85.625^2 (600) + [2(0.5)\pi\,(85.625)(3)](600)$$

$$= 3.93 \times 10^6$$

$$W_{m2} = \pi bGy$$

$$= \pi(0.5)(85.625)(10,000)$$

$$= 1.3 \times 10^6$$

$$W_{total} = 39,250D^2 \text{ (number of bolts) } \frac{30,000}{50,000}$$

$$= 39,250(1.375^2)(88) \frac{30,000}{50,000}$$

$$= 3.91 \times 10^6$$

This joint shouldn't have leaked. The cause was loose bolts.

Increasing the bolt stress to 50,000 lb/in^2 eliminated the leak and could be done because the vessel had a heavy flange that would not distort.

$$W_{total} = 3.91 \times 10^6 \left(\frac{50,000}{30,000} \right) = 6.52 \times 10^6$$

$$> W_{m1} \qquad \text{so acceptable}$$

For additional flange troubleshooting techniques, see the case history in Section 10.6.4.

2.33 CASE HISTORY: HEAT EXCHANGER LEAK DUE TO TEMPERATURE

This case history is concerned with leakage from a shell-and-tube heat exchanger at the tube sheet-to-flange joint. It is a floating-head exchanger because of the high temperature, so there is no stressing of the bundle from the tubes. Discussions with operations personnel and a review of the files revealed the following:

- The leakage appears to have started after rerating the exchanger, when the tube-side temperature increased.
- Leaks seem to coincide with system upsets.
- This is one of several stacked exchangers; at lower temperatures other identical exchangers do not leak.
- The gasket type and bolt torquing are controlled and have not been changed.
- The operating pressure on the tube side is much higher than the pressure on the shell side.

It appears that something is happening to gasket loading at the sealing surfaces and that the higher temperature affects it. Cycling the temperature seems to aggravate the leakage. A useful model to develop would be one that shows the effect of the temperature increase on the sealing ability of the gasket. Such a model will

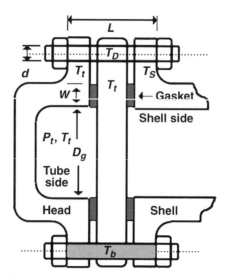

FIGURE 2.65   Heat exchanger head gasket leak due to temperature.

require the clamping force on the gasket to be determined. So the purpose of the analytical model was to determine the most probable cause of the leakage and to recommend a plan to eliminate it.

Figure 2.65 shows the important parameters used in the analysis. In the model only the pressure on the tube side and the tube-side gasket are considered. It is the tube-side gasket that is leaking. Since the tube-side pressure is significantly higher than the pressure on the shell side, the shell side will be considered as unpressured and the tube side will be represented as p. The joint load on the gasket is due to the bolt-up assembly load and the operating load due to the pressure and temperatures.

The bolt load for an assembly bolt stress of σ_a is

$$F_{\text{bolt}} = \frac{\pi}{4} d^2 \sigma_a$$

and the stretch of a bolt due to this load is

$$\delta_{\text{bolt}} = \sigma_a \frac{L}{E}$$

The pressure load acting on the head side is

$$p \frac{\pi}{4} D_g^2$$

and the stretch of the bolts due to this load is

$$\delta_{pressure} = \frac{[p(\pi/4)D_g^2]L}{(\pi/4)d^2NE}$$

$$= \frac{(pD_g^2)L}{d^2NE}$$

The temperature of the tube sheet flange, head flange, and gaskets also affect the bolt stretch. When they are hotter than the bolt, they will try to stretch the bolt and increase the gasket and bolt load. Conversely, when they are colder than the bolt, they will unload the gasket and bolt.

For this analysis a very simplified approach is taken. It assumes that the tube sheet and head flange are at the tube-side temperature on the head-side gasket. It also neglects the shell-side flange and that the tube-side pressure controls. This may or may not be true for other cases but was an acceptable assumption for this case history.

The stretch of the bolts is due to the temperature difference between the bolts and the metal and gasket it holds together:

$$\delta_{temperature} = \alpha L(T_t - T_b)$$

Since the pressure loads the bolt, the resulting stretch in the bolt is

$$\delta_{bolt} + \delta_{pressure} + \delta_{temperature}$$

The bolt stress is

$$\sigma_b = (\delta_{bolt} + \delta_{pressure} + \delta_{temperature})\frac{E}{L}$$

Since the pressure unloads the joint, $\delta_{pressure}$ is subtracted and the resulting joint load is

$$joint\ load = (\delta_{bolt} - \delta_{pressure} + \delta_{temperature})N\left(\frac{\pi}{4}d^2\right)\frac{E}{L}$$

The stress on the gasket surface is

$$\sigma_g = \frac{joint\ load}{gasket\ area}$$

$$= \frac{joint\ load}{w\pi D_g}$$

These equations can be input into a spreadsheet and the gasket and bolt stresses evaluated for various pressure and temperature conditions (Figure 2.66). With too little gasket stress, the joint could leak. With too much bolt stress, the bolt, joint, or

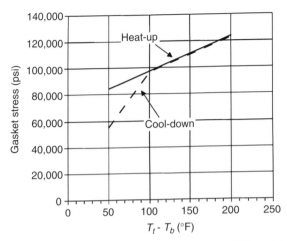

FIGURE 2.66   Gasket stress with temperature differential.

flange could yield. This would result in loss of the elastic nature of the joint so there would be no recovery under temperature or pressure changes and the joint would leak.

Data: 68 1-in.-diameter bolts 11 in. in free length, bolts tightened to 50,000 lb/in²; $D_g = 24$ in.; $w = 0.5$ in.; T_t, T_b, and p data in Table 2.23.

This was a spiral-wound gasket, and it is usually recommended that the gasket stress be less than 44,000 lb/in². This in itself wasn't a concern since the ring controlled the gasket crush. The concern was that with the cycling the gasket might not be able to recover resiliency and thus leak. A different type of gasket, which used grooved metal with both sides filled with graphite tape, was used which could take

TABLE 2.23   Temperature and Pressure Conditions

T_t (°F)	T_b (°F)	p (lb/in²)
	Heat-Up Conditions	
200	150	50
300	200	150
400	250	200
500	300	250
	Cool-Down Conditions	
500	300	250
450	350	250
400	300	200
225	275	50

much higher stresses. This, together with a wider gasket and some procedural heat-up and cool-down changes, helped eliminate the leakage.

In Chapter 10, analytical models to help troubleshooting are discussed in more detail. This case history was included here to show how operational procedures, gasket type and geometry, pressures and temperatures, and bolt-up procedures could all be included in this model. No calculations were included in the model that had not been discussed, and are quite simple.

2.34   EQUIPMENT WEAR

In this book, *wear* is defined as the unwanted removal of material. Not all removal of material is detrimental. In machining operations and other metal-processing operations, often the objective is to maximize material removal but to minimize tool wear. However, in most pieces of equipment wear is usually undesirable and life limiting. Certainly, on bearings, cams, pistons, cylinders, screws, and other surfaces in rubbing contact, wear is to be minimized. Grease and oils are used to lubricate surfaces in contact, and filtering is done to remove abrasive particles from the lubricant. Hard surfacing is used to lower the wear rates of surfaces, as are the use of material combinations that reduce wear and metal transfer by galling.

In this section a simple technique is used to analyze wear. It essentially considers two rubbing surfaces with boundary lubrication between them and a normal load acting on the surface as shown.

Linear Wear Model Development   In this model (Figure 2.67) the sliding distance causing the wear is S, the velocity of the sliding body is V, and the amount of wear is δ, occurring in time t. The wear component, which is assumed to be much softer than the nonwear surface, has a hardness H (kg/mm^2); σ is the nominal pressure between the block and the surface; K, the wear coefficient, has no units; and the "footprint" of the block on the nonwear surface is represented as area A. L is the load acting on the block [23].

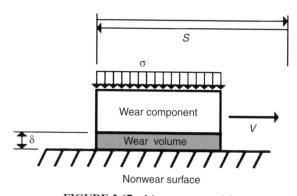

FIGURE 2.67   Linear wear model.

The volume removed by wear is

$$W = A\delta = \frac{KLS}{H}$$

$$S = [V \text{ (in./min)}]t_{min}$$

L/A, the normal stress between the surfaces, is

$$\sigma = \frac{L}{A} \qquad \text{lb/in}^2$$

The hardness H can be converted to Brinell hardness (BHN) units (lb/in^2):

$$H = 1422[\text{BHN (lb/in}^2)]$$

Substituting after converting minutes to years, the design equation becomes

$$t = \frac{(\text{BHN})\delta}{370K\sigma V} \qquad \text{years}$$

Rotational Wear Model Development   Now consider the case of a shaft rotating in a barrel as shown in Figure 2.68. The rotational velocity in terms of rpm is

$$V = \pi[D \text{ (in./rev)}](\text{rpm}) \qquad \text{in./min}$$

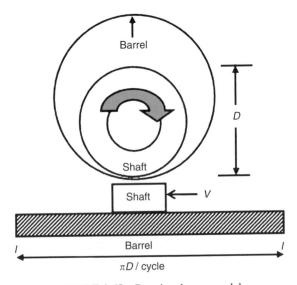

FIGURE 2.68   Rotational wear model.

Making the substitution for V results in the rotational wear equation:

$$t = \frac{(BHN)\delta}{1162K\sigma D(\text{rpm})} \quad \text{years}$$

Looking at the rotational wear sketch, it is evident that a zone on the shaft rubs the barrel only once per revolution. In actuality, the shaft moves around the barrel as it rotates, as is evident on wear patterns in the barrel of extruders. For this reason the full surface of the shaft is assumed to be in contact during a revolution. Examples of the use of this equation and the meaning of the constants are presented in Section 10.4.

2.35   CASE HISTORY: EXCESSIVE WEAR OF A BALL VALVE

Large ball valves can be used as quick-acting product-handling devices which cycle many times per hour. Leakage due to wear can result in product losses. This is a case where the valve seat has experienced heavy abrasive wear of 0.125 in. in only eight months (Figure 2.69). Historically, life in similar services was two years. In this design the seat stress due to the spring pressure had been increased from 1000 lb/in^2 to 4000 lb/in^2 to promote better sealing. The ball surface was much harder than the seat, and the seat hardness had been increased from 371 BHN to 475 BHN to overcome possible wear due to the higher seat loading. The short life suggests that this was not effective. The manufacturer recommended that the seat stress be reduced to 2000 lb/in^2 and the seat hardness raised to 613 BHN; however, there were no test data to verify that there would be an increase in life.

In the following wear, a model was developed to help validate the life claim. The model considers two rubbing surfaces, with boundary lubrication between them, and a normal load stress acting on the surface as shown in Figure 2.67. The sliding distance

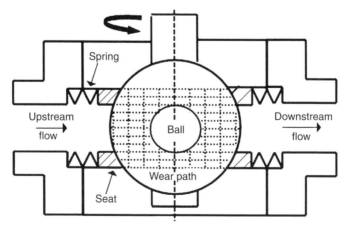

FIGURE 2.69   Ball valve wear model.

causing the wear is S, the velocity of the sliding body is V, and the amount of wear is δ, occurring in time t. The wear component, which is assumed to be much softer than the nonwear surface, has a BHN H; nominal pressure between the block σ; block footprint area surface A; and K, the wear coefficient, has no units.

By converting the geometry and variables of linear surfaces into that of a ball, the time for the surface to wear amount δ in a given time t can be determined. In equation form this is represented as

$$t = \frac{(\text{BHN})\delta}{9.7KDN\sigma} \quad \text{years}$$

where δ is in inches, N is in cycles per hour, σ is in lb/in^2, and D is in inches.

Since a wear life is known, the problem becomes quite simple, and only ratios using the wear equation need be considered. This is because D, N, δ, and K are constant since the same valve is being analyzed; only the stress and hardness change. Using the base-case information of two-year life at 1000 lb/in^2 and 371 BHN and the "compromise" case of 2000 lb/in^2 at 613 BHN, we obtain

$$\text{life (years)} = 2\left(\frac{1000}{2000}\right)\left(\frac{613}{371}\right) = 1.65 \text{ years}$$

From this it appears that the manufacturer's recommendation should provide for a longer life. This conclusion was implemented and ran successfully for over three years.

The analytical model was adequate here since a comparison with the same material under the same conditions was being made. Great care is needed when changing to another type of hard surfacing. The metallurgy, bonding method, and compatibility with the mating surface and product must be taken into consideration. Controlled wear testing and discussions with a materials specialist are always recommended when considering changes in a hard surface.

3

VIBRATION ANALYSIS

3.1 SPRING–MASS SYSTEM AND RESONANCE

Vibration analysis is used less in mechanical engineering than is strength of materials; however, a lack of understanding can be a very expensive proposition when dealing with machinery. Vibration problems on compressors, motors, and ship systems can cause extensive damage. The key thing about vibration problems is that many can usually be reduced to a very simple system for troubleshooting calculations. Although exact results cannot be expected, a better understanding of the problem can.

We discuss vibrations using a case history study. There is a good reason for this, as many complex machines or structures can be reduced as in the examples. This is especially true when modifications to an existing piece of equipment are being reviewed. This case history is based on the vibration of a 2200-hp steam turbine which had a history of startup and in-service vibration problems. Although much vibration testing was done, this analysis will look at understanding the cause and explaining the nomenclature along the way.

This fairly complex multistage steam turbine will be reduced to a simple system so that the fundamental natural frequency can be determined. Since at this time only the rotating members are of importance, the rotor disks can be combined into one mass and the bearing supports and bearing oil film into springs (Figure 3.1). Now from Section 2.6 the shaft can be represented as a simply supported beam that has a k value P/δ. It can then be added as a spring in series and the shaft eliminated.

Analytical Troubleshooting of Process Machinery and Pressure Vessels: Including Real-World Case Studies, by Anthony Sofronas
Copyright © 2006 John Wiley & Sons, Inc.

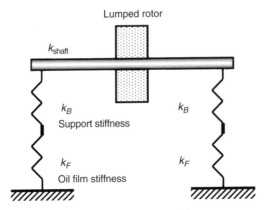

FIGURE 3.1   Simple steam turbine system.

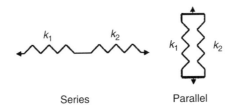

FIGURE 3.2   Springs in series and parallel.

The combined springs use the following equations to obtain an equivalent spring, that is, a spring that has the equivalent spring rate (Figure 3.2).

Springs in series:

$$k = \frac{1}{1/k_1 + 1/k_2}$$

or, in general,

$$\frac{1}{k} = \frac{1}{k_1} + \frac{1}{k_2} + \cdots$$

Springs in parallel:

$$k = k_1 + k_2 + \cdots$$

Combining the springs in the case history and calling the rotor and shaft mass m reduces to the simplified single-degree-of-freedom system shown in Figure 3.3. If the mass m is displaced and then released, or the rotor in the steam turbine is struck,

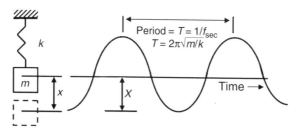

FIGURE 3.3   Single-degree-of-freedom system.

the following motion will occur in an undamped system. The motion of the mass is represented by

$$x = X \sin \omega t$$

For this book we just consider peak values, and by utilizing mathematics arrive at Ref. [24].

$$\text{Displacement } X = X_0 \quad \text{in.}$$
$$\text{Velocity } V = 2\pi f X_0 \quad \text{in./sec}$$
$$\text{Acceleration } a = 4\pi^2 f^2 X_0 \quad \text{in./sec}^2$$

The frequency f represents the number of complete cycles that occur per second. When the frequency is known, the time of a cycle is also known.

$$T = \frac{1}{f} \quad \text{sec}$$

An important fact about this simple single-degree-of-freedom problem is that the frequency at which it will oscillate is simple to calculate:

$$f_n = 9.55 \left(\frac{K}{m} \right)^{1/2} \quad \text{cycles/min}$$

where $m = W/g$.

In review, m is simply the concentrated mass of the rotor and shaft. These are vibrating at displacement X and K is the spring constant, how much the load statically displaces the springs. For a simply supported shaft with the rotor,

$$K = \frac{48EI}{L^3} \quad \text{lb/in.}$$

If the K and m values were determined, it would be a simple task to calculate f_n, the system's natural frequency. This differs from f_f, which is the forcing frequency. The

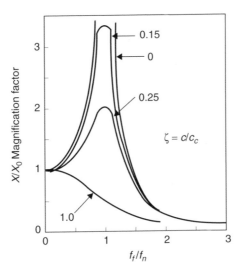

FIGURE 3.4   Magnification factor.

natural frequency is important because it is the frequency at which the system wants to vibrate. If the forcing frequency (e.g., the rotor speed with unbalance) coincides with the natural frequency, resonance will occur. The resonant frequency is important, as high displacements occur with light damping, and $\pm 20\%$ from resonance is a good range to design away from.

Figure 3.4 illustrates how the mass m, displacement X increase the closer to f_f/f_n and is called the *magnification factor*. Also note that damping, represented as ζ, does not greatly affect the frequency, only the amplitude. This shows the importance of calculating this frequency. If the natural frequency of a device is 4500 rpm, which is the same as saying 4500 cycles/min, it would not be wise to have its operating speed range near 3600 to 5400 rpm. Somehow the system should be redesigned for a higher or lower natural frequency, or the speed range changed, or both.

Dynamic loading was discussed in Section 2.10:

$$F = \frac{Wa}{g} = WG\text{'s}$$

If acceleration is now cyclic, based on the displacement curve:

$$G\text{'s} = \frac{a}{g} = 0.051f^2 X_0$$

$$F = 0.051f^2 X_0 W \qquad \text{lb}$$

$$f = \frac{\text{rpm}}{60} = \text{cycles/sec}$$

$$X_0 = \text{in. peak}$$

By knowing the peak displacement X_0 and the frequency along with the weight of the part in vibratory motion, the force to cause this motion can be determined. This can be written as

$$F = 28.4 \times 10^{-6} W\varepsilon(\text{rpm}^2) \qquad \text{lb}$$

where W is in pounds rotating and ε is the eccentricity (in.). This is the same equation as that used in the flywheel problem. An interesting use was on a motor that was moving 0.0005 in. radially at each bearing journal. Was this excessive?

A rule of thumb for balancing machines is that the force on each journal due to imbalance should be less than $\frac{1}{20}$ of the rotor weight. Actually, this means 10% of the rotor weight for allowable imbalance force. Thus, if the rotor weighed 1000 lb, the imbalance should not exceed the following at the journal:

$$1000(\tfrac{1}{20}) \approx 50 \text{ lb}$$

Solving for ε on this 3600-rpm machine, with W equal to 500 lb, we obtain

$$\varepsilon = 0.00027 \text{ in.}$$

This allowable value is less than 0.0005 in., so the vibration is unacceptable. The imbalance at each journal is 92 lb.

In the example above, the rotor was just taken as a mass in space being moved by an unbalanced force. Do not confuse this with a single-degree mass–spring system.

3.2 CASE HISTORY: CRITICAL SPEED PROBLEM ON A STEAM TURBINE

A speed limitation had been imposed on an old steam turbine due to a critical speed in the operating range. Due to the turbine's age, a new, high-efficiency turbine was to be purchased. Several manufacturers had bid on the new turbine, and the manufacturer of the original turbine came in with the highest efficiency and lowest cost. They cited that a design change that stiffened the bearing supports and increased the shaft diameter slightly moved the critical speed to 6800 rpm. This was a good bit above the original 4800-rpm critical speed. The engineer's task was to verify that this was the correct direction to go in raising the critical speed.

Returning to the lumped rotor, spring, single-degree-of-freedom system shown in Figure 3.1, the simplified system shown in Figure 3.5 was developed. Spring constants were determined from available data and are as follows:

$$K_{\text{shaft}} = \frac{48EI}{L^3} = \frac{48\pi E d^4}{64L^3} = 1.7 \times 10^6 \text{ lb/in.}$$
$$K_{\text{oil film}} = 1.6 \times 10^6 \text{ lb/in.}$$
$$K_{\text{support}} = 1.6 \times 10^6 \text{ lb/in.}$$

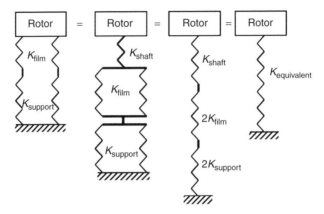

FIGURE 3.5   Steam turbine lumped rotor system.

Solving for the equivalent spring yields

$$\frac{1}{K_{equivalent}} = \frac{1}{K_{shaft}} + \frac{1}{2K_{oil\,film}} + \frac{1}{2K_{support}}$$

$$= \frac{1}{1.7 \times 10^6} + \frac{1}{2(1.6 \times 10^6)} + \frac{1}{2(1.6 \times 10^6)}$$

$$K_{equivalent} = 824{,}000 \text{ lb/in.}$$

Now if the manufacturer made the support very stiff, which was possible, would this raise the critical speed?

If $K_{support}$ were very large, $1/K_{support} \approx 0$ and $K_{equivalent} \approx 1.11 \times 10^6$ lb/in. Indeed, this will raise the stiffness, but what will it do to the critical speed? Recalling the single mass system, we have

$$f_n = 9.55 \left(\frac{K}{m}\right)^{1/2} \qquad \text{cycles/min}$$

Since the mass does not change, the frequency can be proportioned:

$$f_n = 4800 \frac{(1.11 \times 10^6)^{1/2}}{(0.824 \times 10^6)^{1/2}}$$

$$= 5571 \text{ cycles/min}$$

This shows that stiffening the bearing support could be expected to raise the critical speed substantially. Also, the simplified equation for K_{shaft} suggests that increasing the shaft diameter, d, slightly could greatly affect the spring constant and thus the critical speed. Both of these changes were made on the new turbine, and the critical speed occurred at 6800 cycles/min, well away from the 4800-rpm operating speed.

The importance of the example is in the simplification of a complex machine. This allowed the engineer to understand the system and the relative effects of changes. The analysis took one day of an engineer's time, increased his knowledge of the machine, and allowed him the opportunity to ask for specific calculations from the manufacturer. The system has been operating at 5000 rpm for 10 years with no vibration problems.

3.3 DETERMINING VIBRATION AMPLITUDES

In earlier sections, the approach has been to keep the operating frequencies well away from the resonance frequencies. This is a good plan and usually the safest. Unfortunately, this is not always possible, and it may be necessary to go through a critical speed or operate on one. Variable-speed motors, startup of compressor trains, and ship propeller/gearbox systems on ships are examples where operation on a critical speed is possible. Analysis of a forced, damped vibration is well established, and there are many ways to calculate the actual amplitude of vibration at and near resonance. All methods require knowledge of the forcing function and system damping. Since this book considers simple systems (i.e., one-mass, one-spring, single-degree-of-freedom problems), it seems appropriate to show a simplified method to analyze such a system for amplitudes on and off resonance [25].

The major strength of the method is its simplicity, as it allows the resonance amplitude to be estimated with a minimum amount of information. It allows the engineer to use field data to establish a dynamic magnifier number for later use. To use the method, the system must be represented by a spring and a mass, and an exciting torque or force must be acting on the mass. The method is based on knowing the value of the dynamic magnifier M, which is simply the dynamic amplitude divided by the static amplitude. M values are established by historical data available in literature and by experimental testing. This is why the method is only approximate, as the wide range of M values may make the results unusable. The method can produce upper limit amplitudes, since if nothing is known about the damping in a system, hysteresis damping of the steel itself can be used. This produces results of less than infinity, and with a small forcing function the method can produce usable results.

To explain the method, the procedure will be outlined and then applied to some actual systems.

1. Determine the exciting force or torque. Some sources are:

Linear System Source	Linear System Magnitude
Unbalance excitation force	$28.4WR(\text{rpm}/1000)^2$
Piping pulsation	$0.05 \times$ mean pressure
Propeller thrust fluctuations	$0.1 \times$ mean force

Torsional System Source	Torsional System Magnitude
Poor gears, excessive eccentricity	$0.01 \times$ mean torque
Propeller excitation torque	$0.1 \times$ mean torque

2. Determine the static deflection of the simple system:

$$\text{static deflection } Y = \frac{\text{excitation force or torque}}{\text{spring constant } K}$$

3. Determine the magnification factor, M (Table 3.1). When there are several magnifiers, they can be combined into an effective multiplier using the equation

$$\frac{1}{M_{\text{effective}}^2} = \frac{1}{M_1^2} + \frac{1}{M_2^2} + \cdots$$

4. Determine the deflection X at resonance. The magnification factor is defined as

$$M = \frac{\text{resonance deflection } X}{\text{static deflection } Y}$$

The deflection at resonance is therefore $X = MY$.

5. Determine the force at resonance:

$$F = KX \qquad \text{lb}$$

For torsional systems, torsional spring constants, torques, and angles would be used.

TABLE 3.1   Values of the Dynamic Magnifier

	Dynamic Magnifier, M
Torsional system source	
Propeller damping only, ship system	15
Shafts with material damping only	250
Shaft damping with rubber coupling	10
Engines with tuned viscous dampers	10
Engines with no dampers	50
Linear systems	
Bending shafts in hydro-dynamic bearings	10
Piping systems	50
Bending shafts	50

6. The resonance curve away from resonance can be sketched in using the following equation for off-resonance flank amplitudes X_F.

$$X_F = \frac{Y}{[(1 - r^2)^2 + r^2/M^2]^{1/2}}$$

where r = operating frequency/resonance frequency.

3.3.1 Allowable Levels for X or F at Resonance

What is allowable should always be defined by the appropriate codes when available. Some rough rules that are sometimes used are as follows:

- For driven accessories, keep the vibratory angle below $\pm \frac{1}{2}$ degree.
- For vibratory torque on gearing, keep below one-third of mean torque.
- For piping low frequency vibration keep below 0.01 in. peak to peak.
- For shafts in torsion, keep vibratory stress below ± 3000 lb/in.^{2}.
- To keep cracks from growing in steel structures, keep vibratory stress below ± 2000 lb/in.^{2}.
- For machinery, keep reaction forces due to vibratory forces below $0.05 \times$ machine weight.

3.4 CASE HISTORY: VIBRATORY TORQUE ON THE GEAR OF A SHIP SYSTEM

One very important reason for estimating the amplitude of vibration in a geared system is to determine if *hammering* of the gear teeth is possible. Hammering is a phenomenon in which the vibratory torque on the gear is greater than the mean torque on the gear. Vibratory torque allows the gears to separate and strike the front and back of the gear teeth. This occurs at the critical speed frequency and can result in rapid gear failure. In many ship systems the propeller and propeller shaft act as a simple spring–mass system. It is usually not possible to avoid operation at a critical speed, so determining the actual amplitude of vibration along with the critical speed is important. In this example we examine the vibratory torque on the gearing of a seagoing tug.

Figure 3.6 shows the simplified system and its reduction to an equivalent system:

$$K = 18.3 \times 10^6 \text{ in.-lb/rad}$$

$$J = 3047 \text{ lb-in.-sec}^2$$

$$\text{Critical speed} = 9.55 \left(\frac{K}{J} \right)^{1/2} = 740 \text{ rpm}$$

$$\text{Torque at 1000 rpm} = 140,000 \text{ in.-lb}$$

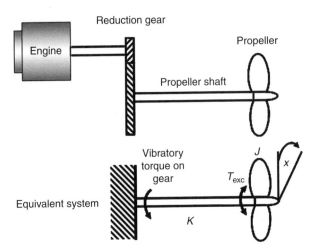

FIGURE 3.6 Ship equivalent torsional system.

Ship torque demand curve:

$$\text{Torque} = 140,000\left(\frac{\text{rpm}}{1000}\right)^2$$

Mean torque at resonance = 77,000 in.-lb

$$T_{\text{exc}} = 0.1(77,000) = 7700 \text{ in.-lb}$$
$$Y = \frac{T_{\text{exc}}}{K} = 0.0004 \text{ rad}$$
$$M = 15$$
$$X = MY = 0.0063 \text{ rad}$$

Vibratory torque at resonance = KX = 115,500 in.-lb

The flank amplitude is determined by the equation in step 6, Section 3.3.

Figure 3.7 represents the ship system before it was tuned by addition of a torsionally flexible coupling. Notice that not only is the resonant peak amplitude shown on the graph, but also the flanks of the resonance curve. It is important to note that the gears will separate because the vibratory torque is greater than the mean torque transmitted through the gearing to the propeller output shaft. The system has to be redesigned. The vibratory torque should be less than one-third of the mean torque, which it certainly isn't.

Systems are generally quite complex, and there are computer programs that handle 100 or more spring–mass systems, with branches, and are available for forced vibration analysis. In practice, many problems can be better understood by simplifying them first, doing some quick calculations, and comparing the solutions with the computer solutions. In the example just discussed, the actual system consisted of

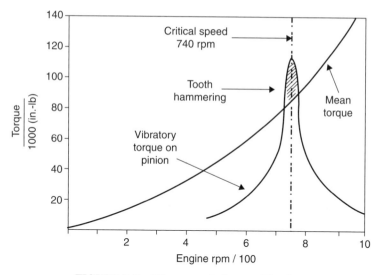

FIGURE 3.7   Ship system before modification.

over 40 masses; however, the first mode natural frequency was within 5% of the simplified model. As the exciting torque data and damping data were not known, the detailed computer model for vibratory torque was no better than the simplified model.

3.5   TORSIONAL VIBRATION

As in the ship system, rotating systems can vibrate when the rotational speed is the same or near the natural frequency of the system. This resonance can be avoided by changing the operating speed. If that is not possible, you have to alter or tune the system inertia's and shaft stiffness, either individually or in combination. Tuning can be difficult for systems with several masses, but the analysis of many systems can be simplified if rigidly connected masses are lumped together and combined into a two-mass system. This allows a good approximation of the first mode frequency.

The natural frequency of a two-mass system in equation form is

$$f_N = 9.55 \left(\frac{J_L + J_S}{J_L J_S} C \right)^{1/2} \quad \text{cycles/min}$$

J_L is the larger torsional mass inertia (lb-in.-sec^2), J_S the smaller torsional mass inertia (lb-in.-sec^2), and C the torsional stiffness of connecting shaft (lb-in./rad).

The critical speed is determined as

$$\text{critical speed} = \frac{f_N}{N}$$

where N is the number of impulses per cycle, such as power piston strokes or imbal-ance. If the critical speed is within the operating range, the value of C, J_L or J_S must be altered. Usually, C is the only value that can be changed readily. When the shaft is steel, the equation for C is

$$C = \frac{1.18 \times 10^6 (D^4 - d^4)}{L} \qquad \text{in.-lb/rad}$$

where D and d are the outside and inside diameters (in.) and L is the shaft length (in.). Since L is usually a fixed length, the diameters, usually D are changed to alter the critical speed.

When several shafts of various diameters and length are bolted together, the equivalent stiffness is used, as in the linear cases:

$$\frac{1}{C_{\text{equivalent}}} = \frac{1}{C_1} + \frac{1}{C_2} + \cdots$$

With these equations you can modify a system so that the critical speed falls outside the operating speed.

3.6   CASE HISTORY: TORSIONAL VIBRATION OF A MOTOR–GENERATOR–BLOWER

A three-mass motor–generator–blower system is designed to operate at 1800 rpm (Figure 3.8). The motor is connected to the generator by a 5-in.-diameter solid shaft 25 in. long. Assume a one-pulse-per-cycle (first-order) excitation due to imbalance. Determine if there is a problem.

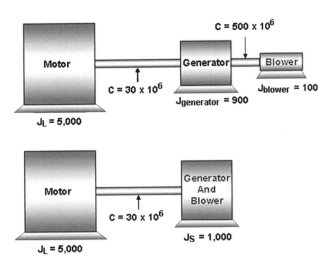

FIGURE 3.8   Motor–generator–blower simplification.

Because of the high stiffness level between the generator and the blower, and the low mass inertia of the blower, the blower can be combined with the generator for the first mode calculation. A rule of thumb is that the end masses can be lumped if the adjoining mass (generator) is five times or more that of the end mass (blower), and the end mass spring is five times or more that of the adjoining spring [26]. Combining the mass inertia's $J_{generator}$ and J_{blower} produces a value of 1000 for J_S.

$$C = 1.18 \times 10^6 \left(\frac{5^4 - 0^4}{25} \right) = 30 \times 10^6 \text{ in.-lb/rad}$$

$$f_N = 9.55 \left(\frac{J_L + J_S}{J_L J_S} C \right)^{1/2} \quad \text{cycles/min}$$

$$= 9.55 \left[\frac{5000 + 1000}{5000(1000)} (30 \times 10^6) \right]^{1/2}$$

$$= 1811 \text{ cycles/min}$$

The critical speed is $f_N/N = 1811/1 = 1811$ rpm. Since the operating speed is 1800 rpm and the critical vibration is 1811 cycles/min, the one cycle pulse per revolution will put it in resonance.

The shaft diameter could be increased from 5 in. to 6 in.

$$C = 1.18 \times 10^6 \left(\frac{6^4 - 0^4}{25} \right) = 62 \times 10^6 \text{ in.-lb/rad}$$

$$f_N = 9.55 \left[\frac{5000 + 1000}{5000(1000)} (62 \times 10^6) \right]^{1/2}$$

$$= 2605 \text{ cycles/min or 2605 rpm}$$

This value is 45% above the operating speed of 1800 rpm and the system can be operated without excessive vibration. Also, it will not go through the critical speed on startup, and the shaft stresses will be low since the diameter was increased. This is a good solution.

3.7 VIBRATION DIAGNOSIS AND CAMPBELL DIAGRAMS

The frequencies associated with a vibration signal are sometimes known. This would be the case if a spectrum analyzer were used to collect the vibration data. The total vibration magnitude would be reduced to the vibration harmonic that makes up the trace. For example, a 0.6-in./sec vibration might consist of 0.4 in./sec at running speed ($1 \times$ rpm), 0.15 at two times running speed ($2 \times$ rpm), 0.05 in./sec at $3 \times$ rpm, and nothing any higher. All of this is important information, as each of these frequencies is associated with a specific part of the machine system, and changes can indicate a problem area.

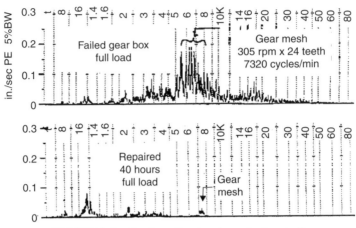

FIGURE 3.9   Gearbox vibration due to broken teeth.

This is especially true if an increasing trend is occurring in one of the components.

The two traces shown in Figure 3.9 illustrate this point by showing the increase in the vibration on a gearbox when the gear has broken teeth. Note that the frequency is that of the gear meshing frequency, which is simply the shaft speed in rpm times the number of teeth on the gear. It is important to have some baseline value with a well-running machine, but in this case a problem is obvious.

Even when data are not being taken and only a system's resonant speeds are known, some knowledge of possible excitations is important. The following should be helpful:

Excitation Frequency	Possible Source
$1 \times$ rpm	Unbalance, misalignment
$2 \times$ rpm	Misalignment
$n \times$ rpm	Problem could be associated with n blades, n gear teeth, n balls or rollers in a bearing, n cylinders on a pump

When one of these frequencies coincides with a calculated or actual critical speed, problems can be expected. The vibration examples show how to remedy some of these problems.

A very convenient method of displaying possible vibration problems is on a Campbell diagram, named after Joseph Campbell, who determined in the 1950s that it was a good method to present turbine blade information. The diagram shows graphically where possible excitation sources intersect with system critical speeds, either lateral or torsional. Construction of a Campbell diagram is quite simple and consists of drawing the rpm and cycles/min axis to cover the range of interest. The $1 \times$ rpm, $2 \times$ rpm, and so on, lines are drawn as shown in Figure 3.10. Horizontal lines are drawn at the critical speeds calculated. Intersections of the lines indicate

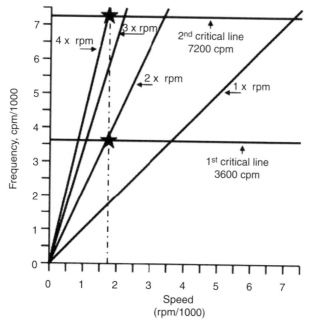

FIGURE 3.10   Campbell diagram.

points at which the critical speed could be excited. The example shown represents a 1800-rpm electric motor–driven pumping system. The system has several natural frequencies or critical speeds since it is a multimass system. The first critical speed is 3600 cycles/min and the second is 7200 cycles/min. The motor is fixed in speed at 1800 rpm.

Notice that at 1800 rpm the first critical speed will be excited by a 2 × rpm disturbance, and the second critical speed will be disturbed by a 4 × rpm disturbance. This is not a good situation, since a slight amount of misalignment could excite the 3600-cycles/min critical speed. The 7200-cycles/min speed would need to be investigated for items that could cause 4-cycle/revolution disturbances before it became a concern.

Only two critical speeds are shown in this simple example, but the usefulness of such a display for more complex systems should be evident. Internal combustion engines of the four-cycle variety, such as used in automobiles, also have $\frac{1}{2}$ orders to worry about: that is, $\frac{1}{2} \times$, $1 \times$, $1\frac{1}{2} \times$, $2 \times$, $2\frac{1}{2} \times$,.... This is due to the four strokes per cycle inherent in the design. The diagram becomes extremely useful in such situations, especially if the speed varies.

Many instruments are available to measure vibratory amplitudes and frequencies. The concern of the engineer is to determine if the values measured are acceptable. Figure 3.11 is a useful guideline for linear vibrations of machines rigidly mounted to foundations. It is not for equipment mounted on isolators, such as rubber automotive engine mounts, which are designed for large motion so that they can isolate a struc-

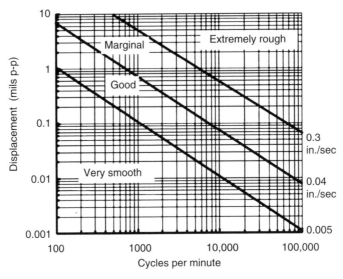

FIGURE 3.11   Machine structure vibration chart.

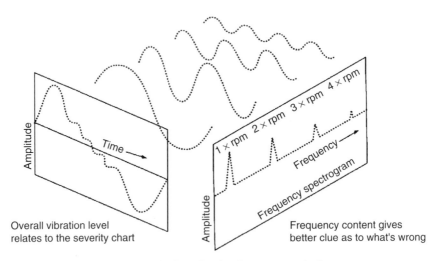

FIGURE 3.12   Vibration frequency analysis.

ture from vibration. Machines and structures that are operating above the marginal velocity line are suspect. Remember that baseline data on amplitudes and frequencies are for a machine operating well, and deviations from this are the best indicators.

Spectrum analysis is a way to break down a vibration signal into its harmonics of various waveforms and their associated amplitudes. Figure 3.12 illustrates how the vibration signal is reduced electronically to its individual components. The benefit of monitoring vibration signals in this manner is that the method can be used to trou-

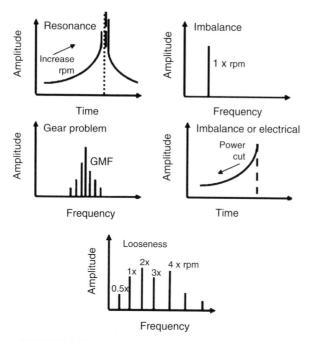

FIGURE 3.13   Common vibration spectrum signatures.

bleshoot a problem without bringing down the machine. With a credible logic scheme, it can also be used as an emergency shutdown system. As mentioned, various failure events have associated frequency signatures. For example, imbalance is usually associated with a 1 × shaft rpm. Some of the more common signatures are shown and explained in Figure 3.13.

In the resonance case, as the speed is increased, the amplitude will gradually increase, and when it goes through a resonance it will increase dramatically. As it passes through, the amplitude drops off just as quickly. When there are higher modes, such as the second or above, the amplitude will do the same again as it increases in speed. The same will occur when the speed is being reduced. If the power is cut, and during coast-down the same type of trace occurs, there is good reason to believe that it is a resonance peak. Resonance is accompanied by a phase angle change.

Imbalance usually manifests itself as 1 × rpm. This is because as the rotor goes through one revolution (i.e., one cycle), the imbalance occurs once. The curve or electrical imbalance is usually of importance on 3600-rpm motors. Since this could be false 60-Hz electrical noise, a check that is sometimes used is to cut the electrical power and watch the coast-down. An electrical problem would disappear immediately, as the dashed line shows. Since in Section 2.10 we saw that imbalance is a function of the square of the speed, an imbalance condition should appear as an exponential curve as shown in Figure 3.13. Obviously, if a resonance

TABLE 3.2   Vibration Frequencies

Equipment	Frequency
Fans	Number of fan blades $\times$ rpm
Impellers	Number of impeller blades $\times$ rpm
Gears	Number of teeth $\times$ rpm
Bearing defect inner race (approximate)	$0.6 \times$ number of balls or rollers $\times$ rpm
Bearing defect outer race (approximate)	$0.4 \times$ number of balls or rollers $\times$ rpm
Bowed shaft	Same as imbalance, $1 \times$ rpm

peak occurs on the way down, the imbalance might have excited a synchronous vibration.

The gear trace was also shown in this chapter when the gear teeth had been ruined. The predominant frequency is the gear meshing frequency, with much "hash" or sidebands on each side. The gear mesh frequency is simply the shaft rpm times the number of gear teeth on the gear on that shaft. Aerodynamic, fluid dynamic, and other signatures are much like that of the gear signatures. The vibration frequency is simply as shown in Table 3.2.

Looseness shows an abundance of frequencies of similar amplitudes. This comes from the manner in which the spectrum analyzer tries to fit functions to impacts that occur during looseness and are not really associated with a given component. This type of trace was observed during work on pump/motor systems when a soft foot occurred. *Soft foot* is a term used when one mounting point or foot of a motor or pump pulls down more than the others. This introduces a misalignment that shows up principally as $1 \times$ but may also have harmonics. Loosening and retightening each mounting point one by one will indicate such a problem. The vibration will be reduced on the soft foot. A dial indicator on the foot will also reveal such a problem. The reading will change by a few thousandths of an inch.

3.8   CASE HISTORY: EFFECT OF A TORSIONAL LOAD APPLIED SUDDENLY

This case history is an actual example of debris passing through a screw-type compressor system. The rotor, which is the first mass, experienced a sudden 42,000-in.-lb torque spike that lasted 0.025 sec. With this sudden spike it was desired to determine the dynamic torque induced in the shaft. The two-mass system shown in Figure 3.14 represents a simplified five-mass system. As the two adjacent shafts had very high spring constants, three masses could be combined into one. The purpose of this section is not to present a design tool but to show the results of an analysis so that the reader can better understand what happens to a system under torsional impact.

The transient response (Figure 3.15) was calculated using a finite-element code

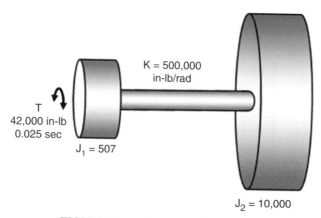

FIGURE 3.14 Two-mass sudden impact.

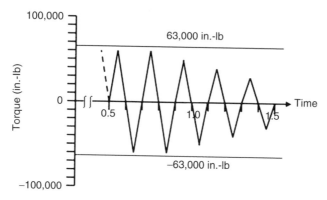

FIGURE 3.15 Transient response in a two-mass system.

and indicates that after the initial impact, the shaft torque reaches a value of 1.5 times the input torque:

$$\frac{63{,}000}{42{,}000} = 1.5$$

It then starts to decay due to system damping. With no damping the amplitude would not go to infinity as in the case of forced vibration. This is because a continuous source of input energy is not available to allow the amplitude to build. The frequency at which the system vibrates is the natural frequency of the system and can be approximated from the plot as follows:

$$\text{frequency} = \frac{\text{number of cycles}}{\text{time (sec)}}$$

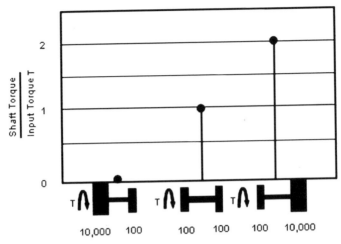

FIGURE 3.16   Sensitivity analysis of a two-mass system.

The reader can prove this by using the equation for a two-mass torsional system and comparing the value calculated with the value above.

This is a very useful method for determining the natural frequency of a structure. With the correct vibration monitoring instrumentation available, the system is struck with a soft mallet when at rest and will ring at its natural frequency. But care must be taken with this approach. Once the author rang the shaft of an overhung pump impeller to determine its natural frequency, and instead, determined the natural frequency of the rather flexible steel bench on which it was bolted.

Figure 3.16 is a sensitivity analysis that illustrates the effect of the masses on the shaft torque. Notice how the shaft torque is attenuated with the larger input inertia. This is understandable when you think of extremes. An infinite inertia would not allow any rotation of the first or second mass. Since the shaft torque is simply the product of the shaft twist and spring constant, there would be no torque in the shaft. Only the dynamic response was determined in this analysis. It should be understood that if there is a mean torque in the shaft, the vibratory torque needs to be added to it algebraically.

From this example one might make the generalization that a factor of 2 on the mean torque of a torsional system would be adequate to handle any transients. This would be a dangerous assumption. When large synchronous motors are the drivers, they can develop air gap torques seven to 10 times greater than the mean torque. This usually occurs when a voltage drop and then a restart attempt occur in too short an interval. In one case the author witnessed a transient of five times mean torque, which caused coupling bolts to fail. A time-delay relay was wired incorrectly and did not function, and the motor restarted too soon after trip-out.

3.9   FLOW-INDUCED VIBRATIONS

In flow over cylinders or similar geometries, vortices can develop on the opposite side from which the flow approaches. Called *vortex shedding*, this is important because it can result in destructive amplitudes on bridges, offshore pipelines, wires, towers, and other structures. These vortex shedding, frequencies are a function of the Reynolds number (Re) and the Strouhal number [27]. A simplified procedure for determining and avoiding this phenomenon is as follows:

1. Calculate the Reynolds number:

$$\text{Re} = \frac{UD}{v}$$

where U is the fluid approach velocity (ft/sec), D the cylinder diameter (ft), and v the kinematic viscosity (ft²/sec).

$$\text{Air:} \quad v = 2 \times 10^{-4}$$

$$\text{Water:} \quad v = 1 \times 10^{-5}$$

2. Calculate the vortex shedding frequency:

$$f_S = \frac{SU}{D} \quad \text{cycles/sec}$$

Values for Strouhal S:

Re	S
200 to 200,000	0.2
200,000 to 4×10^6	Varies from 0.2 to 0.5
4×10^6 to 1×10^7	0.3

3. Keep f_S below the system's natural frequency. See Chapter 2 for typical beam natural frequencies. Usually, the ratio f_S/f_N should be less than 0.5.

3.10 CASE HISTORY: HEAT EXCHANGER TUBE VIBRATION

A tube in a heat exchanger has a 1-in.-outside-diameter tube and is 3 ft long between supports. Water is flowing over it at 20 ft/sec. Assume that it is supported on baffles. Is there a problem?

Some additional data:

$$I = 0.025 \text{ in}^4$$

$$E = 30 \times 10^6 \text{ lb/in}^2$$

$$t = 0.083 \text{ in. wall thickness}$$

$$W = 2.4 \text{ lb tube weight}$$

 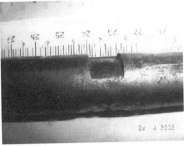

FIGURE 3.17 Tube vibration damage.

$$\text{Re} = 20(1/12)/1 \times 10^{-5} = 167,000$$

$$S = 0.2 \text{ (from table)}$$

$$f_S = \frac{0.2(20)}{1/12} = 48 \text{ cycles/sec}$$

From Section 2.3,

$$f_N = 0.743\left[\frac{(30 \times 10^6)(0.025)}{2.4(3)^3}\right]^{1/2} = 80 \text{ cycles/sec}$$

Since the vortex shedding frequency f_S of 48 cycles/sec is less than the tube natural frequency f_N of 80 cycles/sec in bending, this should not be a problem even though the ratio is 0.6. This is because the support is really between a fixed and a simply supported condition, which would raise the f_N.

Figure 3.17 shows the results when vibration does occur. Some of the tubes to this process exchanger were worn through and leaking, due to the rubbing and clashing of the tubes in the tube baffles. A poor design was the cause.

3.11 CASE HISTORY: PIPING VIBRATION FAILURES

Vibration problems in piping can occur in new installations or in existing systems with abnormal operating conditions. Surging, two-phase flow, unbalanced machinery, fluid pulsations, rapid valve closures, local resonance, and acoustic problems can cause excessive vibration. A new processing plant can have many miles of piping, and the designers use guidelines and experience to know where and how to support the piping. In most new systems it is necessary to "walk the line" to see what's shaking. Vibrating lines are located, marked, and modified by adding gussets, supports, hangers, or snubbers.

Figure 3.18 is based on the author's in-service experience with failed piping. This graph is not a design guide since the database is small. It represents 22 failures, with

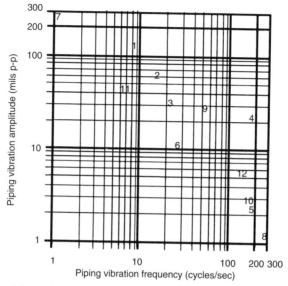

Details on failure points:
1. 24-inch line connected to vessel, 4 ft from weld at vessel
2. 8-inch unsupported line connected to 36-inch header, 3ft from header
3. 12-inch compressor piping failure at support weld
4. 2-inch socket weld 3 ft from socket, no gusseting
5. Crack in weld due to "buzzing" of 2-inch screw compressor line, 2 ft from weld
6. 2-inch pipe to 1/4-inch thin-wall structure, failed at weld to wall, due to wall flexing
7. Reinforced 5-inch branch connection to 8-inch, 4 ft from weld
8. 3/4-inch socket weld, crack at root, lack of fusion, 3 ft from socket weld
9. 8-inch compressor discharge nozzle, 2 ft from weld
10. 6-inch branch connections, flange leaks, weld cracks
11. 1-inch socket weld crack, 2 ft from weld
12. 1/2-inch thread / nipple to gauge failure, in thread, 8 inch from gusset

FIGURE 3.18   Historical piping vibration failures.

several similar failures represented as a single point. It is tempting to draw acceptable/ unacceptable limits on this graph, but one shouldn't. Assuming that no failures will occur in regions where no failures are shown would be a dangerous assumption with this limited amount of data. However, locating a point vibrating at 500 mils peak to peak and 1 cycle/sec with conditions similar to those at point 7 should certainly make one nervous, and it would not be reasonable to expect a long life from this connection. Point 6 was a 2-in. pipe welded to a $\frac{1}{4}$-in.-thick shell with the pipe vibrating. This caused an oil canning effect on the thin metal, and it failed in fatigue in the base metal, not in the weld.

All of these failures could have been avoided by supporting the piping correctly. Good support or gusseting in a vibrating system is mandatory. At a frequency of 15 cycles/sec, well over 1 million fatigue cycles can develop per day. It usually is not a question of if a failure will occur, only when it will occur.

The following case history is based on the failure of a branch connection (Figure 3.19) off the discharge nozzle of a centrifugal pump. The attachments are a block valve, pressure gauge, and bleed-off valve and plug. A technician noticed a

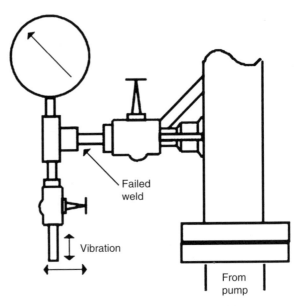

FIGURE 3.19   Gusseted connection failure.

leak at a threaded connection which turned out to be a crack at the thread root. The vibration level during system upsets had been measured as 6 mils peak to peak at 150 cycles/sec at the vibration point shown by the arrows. This is represented as point 12.

It is interesting to note that two-plane gusseting was used in the initial design. The importance of this is that if only one-plane gusseting were used and the vibration happened to occur in the ungusseted plane, vibration in that plane would not be restricted and failures could occur. Unfortunately, a longer nonstandard small-diameter threaded pipe had been added after construction, and it was this connection that failed at the threads. The solution was to strengthened the connection and to seal-weld the threads. This allowed the connection to tolerate upset conditions satisfactorily.

4

FLUID FLOW

4.1 CONTINUITY EQUATION

An important variation in the conservation of mass is the continuity equation:

$$Q = AV \qquad \text{ft}^3/\text{min}$$

It can be stated that

$$Q_{in} = Q_{out}$$
$$A_{in} V_{in} = A_{out} V_{out}$$

When dealing with mass flow rather than the volume flow that was just mentioned, the equation is

$$m = \rho VA \qquad \text{lb/min}$$

For water, $\rho_{water} = 62.4$ lb/ft^3, and for air, $\rho_{air} = 0.08$ lb/ft^3. Multiply this by the specific gravity to get ρ for other fluids. Knowing the density, velocity, and area

Analytical Troubleshooting of Process Machinery and Pressure Vessels: Including Real-World Case Studies, by Anthony Sofronas
Copyright © 2006 John Wiley & Sons, Inc.

flowing in, ρ_1, V_1, and A_1, and the density and area flowing out, ρ_2 and A_2, the velocity out is may be found:

$$V_2 = \frac{\rho_1 V_1 A_1}{\rho_2 A_2}$$

4.2   BERNOULLI'S EQUATION

In the eighteenth century, Daniel Bernoulli investigated the forces present in a moving fluid. This is one of the many forms of Bernoulli's equation and is a very useful form of this energy equation, which is for incompressible flow. Flow is also steady state and frictionless:

$$Z_1 + \frac{V_1^2}{2g} + \frac{p_1}{\rho_1} = Z_2 + \frac{V_2^2}{2g} + \frac{p_2}{\rho_2} + \text{losses}$$

Z is the potential energy term, $V^2/2g$ is the kinetic energy term, and p/ρ is the flow work term. The example in Figure 4.1 illustrates use of the equation and the variables.

An interesting point is that many of the terms in the equation can usually be eliminated by wise choices. Let's determine the flow rate out of the tank due to the head, $Z_1 - Z_2$. This might be because the tank had a hole in it or because someone wanted to deliver a product without a pump.

$$Q_2 = V_2 A_2 \qquad \text{ft}^3/\text{min}$$

The solution can be found quickly by noting that

$$Z_2 = p_1 = p_2 = V_1 = 0$$
$$\rho_1 = \rho_2$$

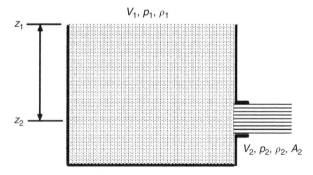

FIGURE 4.1   Bernoulli example.

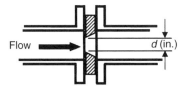

FIGURE 4.2   Sharp-edged orifice flow.

$$V_2 = (2gZ_1)^{1/2}$$
$$Q_2 = A_2(2gZ_1)^{1/2}$$

Although this equation might not be useful to you, it does show how Bernoulli's equation can be manipulated to your specific needs. This equation was once used to determine quickly flow into a water drainage sewer.

4.3   PRESSURE DROP

Another useful form of the equation is flow through a sharp-edged orifice (Figure 4.2):

$$Q_{water} = \begin{cases} 2.5d_{in}^2 \, [\Delta p(lb/in^2)]^{1/2} & ft^3/min \\ 10d_{in}^2 \, [\Delta Z(ft)]^{1/2} & gal/min \end{cases}$$

$$Q_{air} = 70d_{in}^2 \, [\Delta p(lb/in^2)]^{1/2} \qquad ft^3/min$$

4.4   FORCES DUE TO FLUIDS

The following equations are useful if loading due to static or fluids in motion needs to be determined.

$$F = pA$$

For fluids in motion, a form of the momentum equation is used. The force of the fluid stream, shown in Figure 4.3 acting against an area, can be determined by

$$F = \frac{\rho}{g} \, Q(V_{final} - V_{initial})$$

Since $Q = AV$,

$$F = \frac{\rho}{g} \, AV^2 \qquad lb$$

Liquid stream against plate

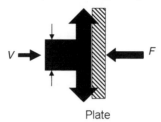

Plate

FIGURE 4.3   Force of a fluid on a plate.

With A in ft^2 and V in ft/sec,

$$F_{air} = 0.0025AV^2$$
$$F_{water} = 1.93AV^2$$

The forces are lower for streamlined bodies.

An interesting use of the equation is to consider the effect of a hurricane wind on a 200-lb person standing broadside in a 100-mph wind. Assume that the person is 1 ft wide and 6 ft tall.

$$F_{air} = 0.005A(\text{mph}^2)$$
$$= 0.005(6)(100^2)$$
$$= 300 \text{ lb}$$

With only 200 lb holding you down, you are about to learn to fly.

The power of a wave coming at you at 50 mph is even more impressive.

$$F_{water} = 4.17A(\text{mph}^2)$$
$$= 4.17(6)(50^2)$$
$$= 63,000 \text{ lb}$$

4.5   CASE HISTORY: PIPING FAILURE DUE TO WATER HAMMER

New plant piping systems designed by piping designers using experience and sophisticated analysis tools rarely fail when operated within the design envelope. Problems occur with add-on systems, which seem to grow over the years. When there is a system upset, poorly restrained piping can move and affect supports or other piping.

In this case we examine a pipe that fell off an overhead rack during such an upset. It broke the support and developed a hydrocarbon leak. The failure investigation team wanted to know what caused the pipe to move and the support to fail. Figure 4.4 shows how the pipe lay on the supports. The vertical riser, valve, and 10-in. line were

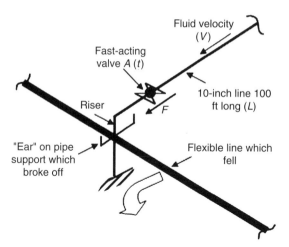

FIGURE 4.4   Piping subjected to water hammer.

added several years after plant construction. During startup, process valve A was suddenly closed after flow was established. Data suggest that the pipe fell out of the rack at this time.

When a valve is closed quickly, a dynamic condition called *water hammer* or *hydraulic surge* can occur. This phenomenon may be noticed when a home faucet is suddenly closed. When the water lines are supported loosely, the pipes may rattle around or a loud thud may be heard. The cause is a pressure wave that can accelerate to the speed of sound in a liquid, due to the velocity change. The resulting pressure differential Δp (lb/in^2) can be determined by considering the change in linear impulse and momentum of the fluid [28]:

$$\Delta p = \frac{0.028VL}{t} \qquad \text{lb/in}^2$$

V is the velocity of fluid in the pipe (ft/sec), L the length of the pipe (ft), and t the time for valve closure (sec).

The average force F (lb) exerted at valve A is due to this pressure differential Δp times the pipe flow area A (in^2). The force is simply the pressure differential times the pipe internal flow area:

$$F = pA = \frac{0.028VLA}{t} \qquad \text{lb}$$

This simplified analysis yields an upper limit on this complex mechanism, and the equation is valid only when the closure time in seconds is greater than $2L/4700$. This is the time it takes the wave traveling at the speed of sound for a round-trip impact at the valve.

For this case, $V = 10$ ft/sec, $L = 100$ ft, $A = 78$ in^2, and $t = 0.25$ sec, which results in a force F equal to 8740 lb, which was enough to cause the "ear" to break. This was confirmed by a metallurgical analysis of the ear, which indicated that several sudden overloads had occurred. It is quite possible that several rapid valve closures occurred over the years, as some of the overload cracks were old.

The effect of the system's flexibility was not considered, and the piping was assumed free to move. Rub marks on the support were used to validate this assumption. A fully restrained pipe would never have hit the support. The equation has limitations and should be used with caution.

The cause was addressed by changing the startup procedure so that the valve would close slowly, thus avoiding the impact type of loading caused by the fast-acting valve. The procedures were changed, in addition to improving the support. Changing only the support could have allowed the pressure pulse to find some other weak point in this complex system.

Here are some areas that an operator can consider to help prevent hydraulic surge or water hammer in piping systems.

- Flow changes in piping should be made slowly, over a period of several seconds.
- Gas systems should be drained properly before startup.
- Liquid systems should be vented properly.
- When possible, pumps should be started against a closed discharge valve and opened slowly.
- When possible, smaller-capacity pumps should be started up before larger capacity pumps.
- When possible, discharge valves should be closed before pumps are stopped.

4.6   CASE HISTORY: CENTRIFUGAL PUMP SYSTEM

In fluid dynamics–related problems, pumps are often involved. The centrifugal pumps used extensively in various process industries are of the type considered in this section. When analyzing a pumping system, three important areas must be considered:

- What the system curve the pump is pumping into and out of looks like
- What the pump performance curve looks like
- The net positive suction head (NPSH) value

4.6.1   System Curves

The entire idea of a pump is to move fluid from point A to point B at certain rates against various pressures (Figure 4.5). A pump introduces a certain amount of energy into a system to do this. Certain losses in the system occur, such as friction through pipes, valves, heat exchangers, or other equipment. Similarly, the pressure in a vessel or level of liquid that must be achieved requires energy.

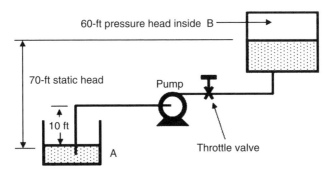

FIGURE 4.5 Pumping system.

The point is that a pump or driver will generate just enough energy to overcome the energy required, and no more. The task then is to determine the system losses. The purpose of the pump is to overcome the static head, pressure head, losses in the piping due to friction, and losses through the throttle valve in the system shown in Figure 4.5. The static head of 70 ft and the pressure head of 60 ft do not change with flow; that is, they are not affected by system friction. The total static head is $70 + 60 = 130$ ft. Here 60 ft is an internal pressure converted to feet:

$$\text{head (ft)} = \frac{[\text{pressure (lb/in}^2)](2.31)}{\text{specific gravity}}$$

Notice that if A were elevated, $-70 + 60 = -10$ ft. This simply means that the pump isn't really needed, as the fluid would flow from A to B by gravity.

From reference books [e.g., 29] on pipe and valve friction losses, the following is obtained with a flow of 1800 gal/min. For a 10-ft length of 6-in.-diameter pipe, the loss is 1 ft, and the flow through a 6-in. valve with a pipe diameter d (in.) and loss coefficient K:

$$h_{6\text{inch}} = \frac{KV^2}{2g} = \frac{0.0026K(\text{gpm}^2)}{d^4}$$

$$= \frac{0.0026(0.12)(1800^2)}{6^4} = 0.8\,\text{ft}$$

The friction loss is small and is only

$$\text{pipe loss} + \text{valve loss} = 1 + 0.8 = 1.8\,\text{ft}$$

This varies as the square of the flows, so at 500 gal/min,

$$h_{500} = \left(\frac{500}{1800}\right)^2 (1.8)$$

TABLE 4.1 System Curve Data

Flow (gal/min)	Friction + Static + Pressure	Sum
0	0 + 70 + 60	130.0
500	0.1 + 70 + 60	130.1
1000	0.6 + 70 + 60	130.6
1500	1.3 + 70 + 60	131.3
2000	2.2 + 70 + 60	132.2

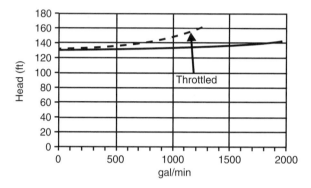

FIGURE 4.6   System curve plot.

This can be used to develop the flow values listed in Table 4.1. The system curve data from the table are plotted in Figure 4.6.

This isn't a good case to illustrate the effects of friction, since they are so small, but the curve can be plotted and the valve throttled to put friction in the system. The dashed curve simulates the system curve with the valve partially throttled. The solid line shows the data calculated, and since the line is almost flat, indicates that there isn't much friction from the pipes, fully opened valves, or other friction losses. Notice the effect of higher flow rates on the system head. Nothing has been said yet about the pump; this represents the system only.

4.6.2 Pump Curves

With the desired system curve developed and the required flow of 1800 gal/min known, the pump performance curves, which are supplied by the pump manufacturers, are matched to the system curve. Normally, it is a good idea to keep the design point within 15% of the pump's best efficiency point (BEP). This saves power and keeps internal forces low. The point is shown on the manufacturer's pump curve. One good way to save yourself some work is to superimpose the system curve onto the pump curve.

Pump curves (Figure 4.7) are busy curves and contain considerable information, such as:

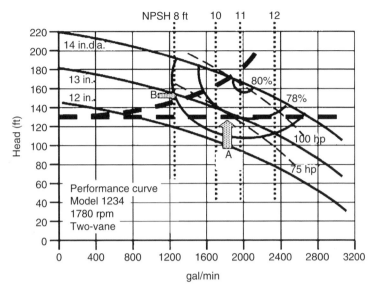

FIGURE 4.7   Pump curve with system curve.

- Head and flow
- Required horsepower
- Required NPSH
- Best efficiency point
- Various-sized impellers

With the curves superimposed, point A represents the operating point at 130 ft and 1800 gal/min. The pump curve in Figure 4.7 for capacity/head is called the *steady-rising* type. It is a stable curve, since a small head rise or drop causes a relatively small flow change.

There are cases where flat curves are required; that is, the curve doesn't rise to the left, but is flat. These types of curves can cause problems in the wrong application, as small head change (i.e., pressure change) can cause large-capacity changes, with the result that no-flow conditions are possible, which is not a good thing. They are useful if you want to maintain constant pressure with variable-flow swings.

From this particular manufacturer's pump curve, A falls on a pump with:

- A 13-in. impeller
- 78% BEP
- 75 hp
- $NPSH_R = 11$ ft

If this pump were selected and installed and the throttle valve partially closed, the pump would operate at point B. Similarly, if less total head than point A were

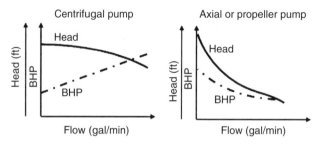

FIGURE 4.8 Centrifugal and axial flow pump curves.

needed, the point would move on the curve to the right and more than 75 hp would be required. This actually happened with this pumping system. The throttle valve was a control valve whose position was determined by a level in a tank. The controller malfunctioned and opened the valve to the full-open position, where it pumped over 3200 gal/min and tripped out the motor on overload.

This example shows the importance of pump curves as problem-solving tools. If the pump were operating at point B, we might be concerned with valve restriction, excessive pressure at the discharge point, or several other problems.

Centrifugal pumps differ significantly from axial-flow or propeller-type pumps, as is evident in Figure 4.8. Although axial-flow pumps are not described in this book, the principles are the same as for centrifugal pumps. In the case history in Section 4.9, a problem describing low flow and high head in a propeller pump is analyzed. In a centrifugal pump, a blocked-in discharge (i.e., a high head) is associated with maximum head and minimum brake horsepower. The axial-flow or propeller pump is the opposite, with maximum head and maximum horsepower occurring in the blocked-in condition. Blocking the discharge will therefore result in higher horse-power for the axial pump and less for the centrifugal pump.

One other very important pump characteristic in selection and troubleshooting is NPSH.

4.6.3 Net Positive Suction Head

Two NPSH terms are important. The first is *NPSH required* ($NPSH_R$). This is the energy needed to fill the pump on the suction side. This information is supplied by the pump manufacturer and is usually on the pump performance curve, as it is a function of flow. The second, *NPSH available* ($NPSH_A$), is a characteristic of the system and a calculated value. It is the energy in a liquid at the pump suction. It must be more than the energy in the liquid due to its vapor pressure, so the liquid does not vaporize in the suction line. A centrifugal pump doesn't pump a gas unless it is of a special design.

The key point is that the NPSH available must always be greater than the NPSH required:

$$NPSH_A > NPSH_R$$

If this is not true, the pump may not pump. The NPSH available equation is

$$\text{NPSH}_A = h_{\text{surface}} - h_{\text{vapor}} \pm h_{\text{static}} - h_{\text{friction}}$$

where h_{surface} is the pressure on the liquid surface (atmospheric pressure is 34 ft), h_{vapor} is the vapor pressure of the liquid (ft; obtain from tables), h_{static} is the static head (ft; minus if lift and positive if flooded suction), and h_{friction} represents all pipe, valve and entrance losses (ft). Remember that h_{surface} would equal h_{vapor} in a closed tank since the vapor generates h_{surface}. Think of a gasoline can left in a hot garage. When you take the top off, there is a slight pressure caused by the vapor pressure of the gasoline. An example of the use of the NPSH_A value in an open vessel is as follows:

$$\text{NPSH}_A = 34 - 1 - 10 - 2 = 21 \text{ ft}$$
$$> \text{NPSH}_R?$$
$$21 > 11 \quad \text{OK}$$

The following are some things that can be done when the NPSH required is not sufficient:

- Reduce the delivery. Splitting the delivery among several pumps, such as using the standby pump, can help.
- Use a larger impeller. NPSH_R will improve, but the power consumption will be worse.
- Operate the pump with cavitation, which may not be severe enough to do damage. Sometimes the manufacturer will agree that the 3% total head delivery head drop can be greater.
- Select a pump with a better NPSH value.
- Change the static head or friction so that more is available.

4.6.4 Pump Laws

Centrifugal pumps generally obey what are known as the *pump laws*. These so-called "laws" allow flows, pressures, horsepower, and impeller diameter to be scaled from one condition to another. Any units can be used as long as they are used consistently.

$$gpm_2 = gpm_1 \frac{rpm_2}{rpm_1}$$

$$H_2 = H_1 \left(\frac{rpm_2}{rpm_1}\right)^2$$

$$hp_2 = hp_1 \left(\frac{rpm_2}{rpm_1}\right)^3$$

Assume that a pump is operating at 1800 rpm, 400 gal/min, at a head of 48 ft and at 50 hp. The pump motor is to be changed to a 3600-rpm motor. What will be the flow, head, and hp at 3600-rpm?

$$gpm_2 = 400 \left(\frac{3600}{1800} \right) = 800 \text{ gal/min}$$

$$H_2 = 48 \left(\frac{3600}{1800} \right)^2 = 192 \text{ ft}$$

$$hp_2 = 50 \left(\frac{3600}{1800} \right)^3 = 400 \text{ hp}$$

That's a big horsepower increase that will take a big motor, coupling, and other components. The diameter of the impeller could be changed for the rpm in these equations if impeller trimming were being considered.

4.6.5 Series and Parallel Pump Operations

In processing plants pumps may be operated in series or in parallel, as shown in Figure 4.9. In *series operation*, discharge from the first pump goes to the suction of the second pump. For two identical pumps, the head will be added and the flow will remain the same. This arrangement is used to increase the discharge head. In *parallel operation*, for identical pumps, the flow rates add but the discharge head remains the same. This arrangement is used to increase the flow, and in this case a common suction is used for both pumps and the discharges are connected together. Such arrangements might be used in pumps for cooling towers, where multiple pumps may be used based on the cooling capacity required.

Problems can occur with this arrangement. Notice in Figure 4.10 that pump 1 produces more than pump 2. This system considers friction only. The pumps could be run at different speeds, or are different sizes, or pump 2 could be worn. Notice that pump 2 will not contribute any flow, and its check valve will remain closed until pump 1 delivers about 1200 gal/min at 120 ft of head. At lower than 120 ft of head, pump 2 starts contributing into the system and adds, as in curve 3.

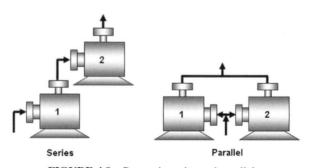

FIGURE 4.9 Pumps in series and parallel.

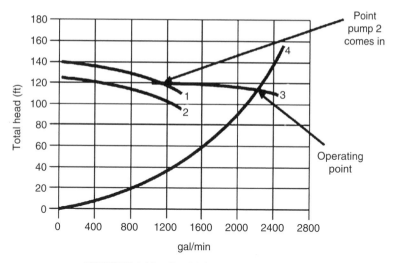

FIGURE 4.10 Combining pumps in parallel.

So when the pumps are different, it is possible for the stronger pump to block out the weaker pump and never allow it to contribute. With several pumps in parallel, some driven by motors and others driven by steam turbines, such as in large cooling towers, this has been a concern and is valuable to remember during troubleshooting.

4.6.6 Blocked-in-Pump Concern

Heat generation can be a problem when a pump is operated against a closed discharge valve. In the system shown in Figure 4.11, the pump will draw 40 hp when blocked in. Since there is no flow, this horsepower will have to be dissipated as heat to the liquid in the case. This will result in a temperature rise. What if the case contained 20 gallons of water or about 200 lb. How much would the water temperature rise?

From the section on heat transfer, we have 1 hp = 42.2 Btu/min, so 40(42.2) = 1688 Btu/min has to be dissipated into the water.

$$Q = WC\Delta T \qquad \text{Btu}$$

The specific heat C of water is 1.0 Btu/lb-°F, $W = 200$ lb, and $Q = 1688$ Btu each minute:

$$\Delta T = \frac{Q}{WC}$$

$$= \frac{1688}{200(1)}$$

$$= 8.5°F$$

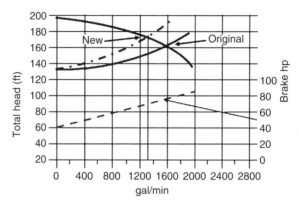

FIGURE 4.11   Blocked-in pump.

So each minute the temperature will rise 8.5°F and must be removed through the case to the atmosphere, piping, or shaft. As you can see, the temperature could build up quite quickly with poor heat dissipation or small volumes of product. Hydrocarbons present other concerns. Blocked in flow can also cause increased radial loads, so should be avoided.

4.6.7   Cryogenic Service Concerns

"Run-dry"-type failures occur more frequently than is desirable with horizontal multistage pumps in liquid ethylene, propane, methane, and other services. Usually, the result has been a wreck on a critical pump, due to contact of the rotating and stationary parts or galling and burning of bushings. Other times, mechanical seals appeared to have run dry. Many of these pumps have been modified to allow them to run dry for short periods of time. Wear rings or bearings made of graphite-impregnated materials and alternative flush arrangements have been used. Karassik [30] provides a good explanation of one possible cause of cryogenic pump failures.

Obviously, it could be just an operating problem and a low-flow condition, but many of the failures have shown the flow to be within normal design. Karassik compares the vapor pressure of water and of methane, at temperature, on a per degree basis.

Liquid	Temperature Change (°F)	Vapor Pressure (ft)
Water	80 to 79°F	0.04
Methane	−240°F to −239°F	8.95

With this type of sensitivity, anything that causes a small temperature variation within one of the pump stages will raise the apparent required NPSH. In water or similar liquids, this would not be a concern, but for many LPG products it could be. The same holds for flashing across a seal face.

Although the use of alternative bushing designs eliminated failures in the horizontal split-line pumps modified, it was probably because it allowed them to run only partially lubricated for a short period of time and could "ride out" the rubbing. It also might have reduced any rubbing that caused the heat. This leads to the question of which came first, the heat or the rub.

4.6.8 Pump Control

A pump can be controlled by several means, and some of the more common are summarized here [31].

- *Discharge valve throttling.* This is similar to restricting discharge of the pump using a valve or orifice and changes the shape of the system curve as shown in Figure 4.6. This will cause it to ride up the pump curve to a new flow and pressure.
- *Suction valve throttling.* This has an effect similar to that of discharge throttling. A major drawback is that of affecting the NPSH, which could result in cavitation or run-dry conditions.
- *Recycle control.* In this method some of the discharge flow is redirected back to the pump suction, and the results depend on the type of curve. With a flat curve, a large change in flow can have very little effect on the process discharge pressure. This might not result in an adequate pressure and flow control arrangement.
- *Speed control.* Variable-speed drive motors are used. This method is not in wide use because of the difficulty in maintaining the process operating point.
- *Riding on the curve.* When you have no control devices at all, the pump will simply follow the pump curve as the system discharge changes. This is the way that most process pumps operate, and although exact flows and pressures are not maintained, the method is adequate for many processes. Some form of minimum flow protection will probably still be required.

4.7   CASE HISTORY: WRECK OF A CENTRIFUGAL PUMP

This example on an NPSH problem may appear simple to solve, but it took over one month of interviews with personnel: data history reviews, plant walk-throughs, schematic reviews, and presentations to an irate plant manager. High-visibility failures are never easy, and a high-reliability plant requires much more than just good troubleshooting and analytical skills [32].

Centrifugal pumps are plentiful in many industries and are usually quite reliable, with only sporadic seal, bearing, motor, and coupling problems. This case involves the destruction of a multistage deep well pump used for transferring a product from a tank to a railcar (Figure 4.12). Three failures had occurred within six months, with the last resulting in a fire. A failure investigation team was formed to determine the cause and how to avoid future failures.

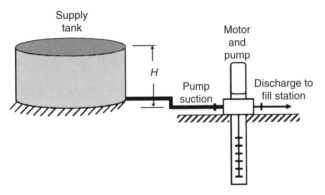

FIGURE 4.12   Centrifugal pump wreck.

During each rebuild of the pump, the assembly was carefully witnessed by plant specialists, and the work was done by a reputable shop. Alignments were carefully documented and verified. The pump had previously run for five years with no problems, so something had changed. The failure evidence, overheating and galling, indicated that the pump had run dry, so the operating conditions were reviewed. As mentioned, a large number of pump system problems are due to inadequate $NPSH_A$ values, which is the amount of head necessary at the suction of a pump to keep the fluid from vaporizing and thus limiting flow. The pump manufacturer provides $NPSH_R$ data or the head required at the suction of the pump at various flow conditions. The system designer is responsible for providing an adequate $NPSH_A$ margin. The cardinal rule is that the $NPSH_A$ has to be greater than $NPSH_R$ or the pump may not pump and cavitation can occur, which can sound like "rattling marbles" in the pump. The equation for $NPSH_A$ is

$$NPSH_A = h_{surface} - h_{vapor} \pm h_{static} - h_{friction}$$

The head h (ft) terms are pressure on the liquid surface, vapor pressure of liquid, static head (minus for lift), and friction losses.

One thing that changed was the minimum level h_{static} in the vessel. It had been lowered to allow more railcars to be filled, thus increasing throughput. The manufacturer stated that $NPSH_R = 15$ ft of head was required at the pumping flow rate being used. For this case, $h_{surface} = h_{vapor}$, $h_{friction} = 10$ ft, $h_{static} = 20$ ft, so $NPSH_A = 10$ ft. Thus, with only 10 ft NPSH available and 15 ft of NPSH required, vaporization and a loss of flow can be expected. The original minimum tank level was 30 ft, so 30 − 10 = 20 ft, which is greater than 15 ft and explains why the pumps ran fine for five years.

Raising the minimum level and de-bottlenecking some of the suction piping friction, $h_{friction}$, eliminated any additional failures. This was a poorly designed system, and although the sketch looks as if is a storage tank open to the atmosphere, it wasn't.

Atmospheric pressure would have made h_{surface} equal to 34 ft, and there wouldn't have been a problem. Also, the minimum tank level was much too high.

4.8 CASE HISTORY: AIRFOIL AERODYNAMIC LOADS

This example is being presented to illustrate the use of lift and drag equations. An approach such as this was used to aid in determining the cause of a structural failure on a drag-racing vehicle (Figure 4.13). There are several forces and moments for which the support bolting and tubing must be designed. The most formidable is the aerodynamic load produced by the airfoil itself. Once this load is known, methods such as those shown in Chapter 2 can be used to determine the load stresses and deflections.

Airfoils are used as wing sections in aircraft because they produce lift with a minimum of drag. The illustration of the dragster shows the airfoil mounted and the definition of the lift and drag forces. The angle is the inclination of the airfoil to the airflow and is called the *angle of attack*. In the figure, the airfoil is shown inverted and the lift pointed down. This is to keep load on the rear tires so that they stay in contact with the race track. If it were installed as a wing section, the rear would try to fly at high speeds. The front end also has an airfoil, but it is not shown. The airfoil is exaggerated in the sketch to show the forces. It is not in correct proportion to the rest of the dragster. Airfoil designs are available through several sources. One convenient source for NACA designs is Ref. [33]. The various charts allow complete design of an airfoil, including the curvature.

The equations for the lift (L) and drag (D) forces should look familiar, as they are similar to the force equation in Section 4.4:

$$L = \begin{cases} 0.0012C_L V^2 S & \text{lb} \\ 0.0026C_L(\text{mph}^2)S & \text{lb} \end{cases}$$

$$D = \begin{cases} 0.0012C_D V^2 S & \text{lb} \\ 0.0026C_D(\text{mph}^2)S & \text{lb} \end{cases}$$

In the equations C_L and C_D are the lift and drag coefficients determined from wind tunnel tests and are dimensionless. They contain the effect of the angle and pressure

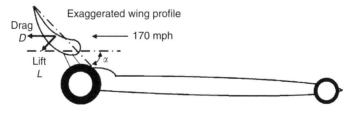

FIGURE 4.13 Dragster wing design.

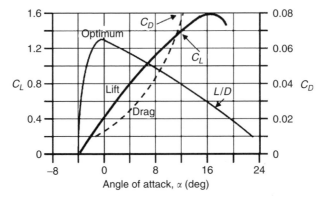

FIGURE 4.14   Airfoil design C_L and C_D.

differentials on the wing as well as skin friction and other losses. V represents the velocity (ft/sec) and S is the surface area of the airfoil (ft^2). Air density is for 70°F at sea level.

Figure 4.14 shows the lift and drag coefficients as functions of the airfoil angle relative to the airflow. Notice that there is an optimum angle for this airfoil since at that point the ratio of lift to drag (L/D) for the design is maximum. For the foil used, the actual angle of attack is 12° when mounted. The S term is the area of the airfoil, that is, the length times the width, in square feet. Since still air is assumed, the speed is simply the vehicle speed.

The dragster has an airfoil 3 ft long and 1 ft wide, or 3 ft^2. With the airfoil at an angle of 12°, the lift coefficient is 1.35 and the drag coefficient is 0.08. At 170 mph the lift (L) and drag (D) forces are as follows:

$$L = 304 \text{ lb}$$
$$D = 18 \text{ lb}$$

The aerodynamic load can now be used to evaluate loads, stresses, and deflections on the frame supports and bolting.

4.9   CASE HISTORY: PRESSURE LOSS THROUGH SLOTS

There are times in an engineer's career when a quick decision is needed so that a design will proceed in the correct direction. There may be no time to undertake a library search or to conduct tests to find a solution to the problem. The condition may be that a modification is being proposed that just doesn't sound acceptable to the engineer, and some information is required to force the decision in another

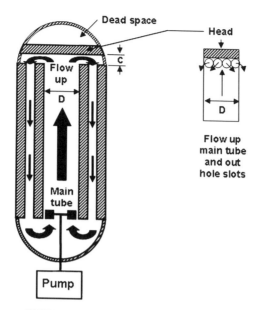

FIGURE 4.15   Reactor head modification.

direction. This is a case where calculations proved the engineer's intuitive feeling that a design change wouldn't work.

In this case history (Figure 4.15), modifications were proposed to a reactor vessel that contained an axial-flow pump. The dead space in the head of the vessel was to be reduced by changing the head design from a dome to a flat head, as shown in the sketch of the circulator. To reduce the volume as much as possible, the clearance between the top of the flow tube and the head was reduced from 8 in. to 1 in. using the plate as shown. On reviewing the drawings this clearance looked like it would result in significant pressure loss to the engineer, and some information was required to prove his point. It should be mentioned that the general opinion of several others who had reviewed the drawing was that it should work. This is the simple analysis the engineer used to change the decision.

Now this is not really a simple problem, so a key item is trying to visualize the problem as something one can solve with the equations and knowledge available. At the time of this analysis the pressure loss through the configuration shown was not known, but the problem could be reduced to a simpler problem for which a solution was known. The loss between a pipe discharging against a flat plate was not known. In the fluid flow section, the solution for flow through an orifice is presented. A pipe with a series of holes around the circumference of diameter D and close together, with water spraying out in a manner similar to a slot, was visualized as shown in Figure 4.15.

Since the solution to the pressure loss through a hole is known, all that really needs to be done is to determine the number of holes the flow will be going through, so that the flow per hole can be determined. The number of holes around the circumference of diameter D is

$$nc = 3.14D$$
$$n = \frac{3.14D}{c}$$

From the fluid flow section, water flow through a sharp-edged orifice produces the following pressure loss:

$$\text{loss} = \left(\frac{Q_h}{10c^2}\right)^2 \quad \text{ft}$$

Since the holes are all the same size, the total pressure drop across all of the holes is the same as the loss through one hole, with a flow of $1/n$ of the total flow through the large tube:

$$Q_h = \frac{Q}{n} = \frac{Qc}{3.14D}$$

Substituting Q for Q_h yields

$$\text{loss} = \left(\frac{Q}{31.4Dc}\right)^2 \quad \text{ft}$$

Here Q is the total flow of water (gal/min), D the pipe diameter (in.), c the clearance or hole diameter (in.), and loss is the pressure drop through c (ft).

The results of the equation for the conditions given are shown in Figure 4.16. From the figure, one thing is evident: The simplified equation shows that the pressure loss

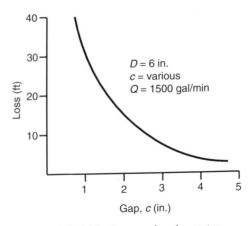

FIGURE 4.16 Pressure loss in reactor.

is going to increase rapidly as the clearance is reduced to less than 3 in. Similarly, the loss is small above 5 in. Some idea of the degree of smallness can be gotten by knowing that the friction of pumping through the tubes alone is 9 ft, and it appears that for a 5-in. clearance c, the loss is less than 2 ft. At 1 in. of clearance the loss is 40 ft.

Pumping at such an elevated pressure is similar to closing a valve on the discharge of a pump. An axial-flow pump is used in this circulator, and from the pump curve, horsepower is maximum at shut-off conditions, as discussed previously. This isn't like a centrifugal pump, where horsepower is minimum at a blocked-in condition. So this much resistance would have been totally unacceptable. One look at the curve and the group agreed that a clearance of at least 4 in. was needed, and the design proceeded accordingly.

This is not an analysis that one should plan on presenting as an exact solution to this problem, since many assumptions were made that were not proven. For example, the area of the holes is less than the area of a slot; also, the pressure drop through a slot is probably going to be less than that of a number of holes. Another thing is that as the clearance c is reduced, at some point friction losses are going to become significant and will reduce the flow. These have not been accounted for, nor have losses due to changes in the velocities direction and magnitude.

The analysis did satisfy its purpose, and later tests showed it to be fairly accurate within the range used. This analysis took two hours of the engineer's time. The wrong clearance would have resulted in lost production time to define what the problem was, as no evaluation would have been available to identify pressure drop as a problem. In addition, $10,000 would have been required to machine and fit a new head, and another $10,000 would have been wasted on the scrapped head design. This does not even consider the production losses on a sold-out unit.

4.10 FRICTION LOSSES IN PIPING SYSTEMS

Following is a simplified procedure for determining the pressure drop through a piping system [29].

1. Calculate the Reynolds number:

$$Re = \frac{VD}{v}$$

where V is the velocity (ft/sec), D the pipe ID (ft), and v the kinematic viscosity (ft^2/sec).
   Some values for v:

Water, benzene	1×10^{-5}
Crude oil	1×10^{-4}
Air	2×10^{-4}
Hydrogen	1×10^{-3}

For Re < 2000, consider the flow laminar. For Re > 4000, consider the flow turbulent.

2. For the case of laminar flow:

$$f = \frac{64}{Re}$$

For the case of turbulent flow (interpolate between values):

Value of D	Value of f	Condition
$\frac{1}{4}$–1	0.040	Commercial pipe
1–5	0.040–0.240	Rough pipe, concrete

3. The pressure drop in the piping is shown in the following equation. The first term is the friction loss due to the pipe length, and the second term added to it is the loss due to valves and other components in the system.

$$\text{Pressure drop} = \frac{(f(L/D) + \Sigma k)V^2}{4.6\,g}\ \text{lb/in}^2$$

where L is the pipe length (ft), k the velocity head loss factor, and $g = 32.2\ \text{ft/sec}^2$. Typical k values:

Globe valve open	10
Angle valve open	5
Swing check valve open	2.5
Standard elbow	1
Gate valve $\frac{1}{4}, \frac{1}{2}, \frac{3}{4}$ open	24, 6, 1

In fluid flow when the process and materials of construction will allow, reasonable guidelines based on friction losses for velocities in pipes from a pressure drop standpoint are

Gases and saturated steam:	less than 100 ft/sec
Liquids:	less than 10 ft/sec

4.11   CASE HISTORY: PIPE FRICTION

Crude oil is flowing through 500 ft of 5.5-in.-ID pipe. Two standard elbows are installed: a swing check valve and a $\frac{3}{4}$-open gate valve. The line is to be purged with nitrogen. What are the pressure drops if crude oil flows at 10 ft/sec? The pipe is commercial quality;

For crude:

$$Re = \frac{10(5.5/12)}{1 \times 10^{-4}}$$

$$= 45,800 \text{ turbulent}$$

$$f = 0.04$$

$$\text{Pressure drop} = \frac{[f(L/D) + \Sigma k]V^2}{4.6\,g}$$

$$= \frac{[0.04(500/0.46) + \Sigma 5.5](10^2)}{4.6(32.2)}$$

$$= 33 \text{ lb/in}^2$$

5

HEAT TRANSFER

In heat transfer steady-state conditions, we are again looking at an energy balance:

$$q_{generated} = q_{dissipated}$$

The heat generated can be from many sources, some of which are discussed later. The heat transfer rate (Btu/hr) or power dissipated ($q_{dissipated}$) is discussed first.

The classic modes of heat transfer are:

- Conduction
- Convection (forced and free)
- Radiation

Each of these modes needs to be understood so that the following relationship can be solved:

$$q_{generated} = q_{conduction} = q_{convection} + q_{radiation}$$

5.1 CONDUCTION

Heat energy is passed from one particle to another and is associated primarily with solids. Holding a hot cup of coffee in an aluminum cup tells you immediately that it

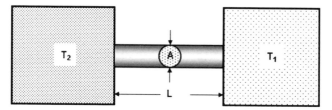

FIGURE 5.1   Heat conduction between vessels.

has a very high thermal conductivity compared to, say, a plastic cup. To understand
the basic *conduction equation*, consider two insulated vessels (Figure 5.1). One is at
a higher temperature, T_2, and the other is at T_1. Both are insulated and connected by
a pipe (solid). Intuitively, we can see that heat will travel through the pipe from the
higher to the lower temperature.

   The equation that quantifies this, *Fourier's law*, is the fundamental law describing heat flow by conduction:

$$q = \frac{KA(T_2 - T_1)}{L} \qquad \text{Btu/hr}$$

The thermal conductivity K values at 100°F for some materials are

$$K_{plastic} = 0.3 \, \text{Btu/hr-ft-°F}$$

$$K_{steel} = 20.0 \, \text{Btu/hr-ft-°F}$$

$$K_{aluminum} = 100.0 \, \text{Btu/hr-ft-°F}$$

This gives us some idea why we can hold the plastic cup with no discomfort, as the
heat travels into our hand at a much slower rate, assuming that A, L, and ΔT remain
the same. As a matter of fact, the rate is

$$\frac{K_{aluminum}}{K_{plastic}} = 333 \, \text{times less}$$

5.2   CONVECTION

In convection, heat energy flows from one location to another by migration of particles from one location to another. Obviously, convection cannot take place in a solid,
but does in liquids and gases. In free convection, heat transfer is by thermal stimu-
lation (heat from a plate in still air). In forced convection some fluid with velocity is
used over the surface. It is a much superior method when applicable. The *convection
equation*, courtesy of Newton (1701), is

$$q = hA(T_s - T) \qquad \text{Btu/hr}$$

At once the similarity between this equation and the conduction equation is apparent. It appears that if h is simply replaced by K/L, the equations will be the same! The problem is that L is now a film thickness and can be different, depending on what type of flow is associated with it, laminar or turbulent. One relation for force convection from large plane surfaces is

$$h = 1 + 0.225V \qquad \text{Btu/hr-ft}^2\text{-}°\text{F}$$

where V is the air velocity (ft/sec) over the plate. When V is zero, the result is roughly that of free convection. This is very crude, and if a further refinement is warranted, a more detailed calculation of h should be obtained.

Another useful forced convection equation considers fluid flow across the surfaces and is a form of the specific heat equation:

$$q = mC(T_2 - T_1) \qquad \text{Btu/min}$$

where m is in lb/min, C is in Btu/lb-°F, and T_1 and T_2 are in °F. In the equation, C is the specific heat of the substance, which is defined as the amount of energy (i.e., heat) required to raise the temperature of a unit weight of a substance $1°$ in temperature. These values are well tabulated for gases, liquids, and solids. For gases, C depends on whether it is a constant pressure, free to expand, or a constant-volume, closed-container process (Table 5.1).

The specific heat equation is also useful in the weight form:

$$Q = W_{lb}C(T_2 - T_1) \qquad \text{Btu}$$

It can be used to determine how much energy is required to raise some mass by ΔT.
One useful form of these equations is the temperature rise across airflow:

$$\Delta T = \frac{3715(\text{kW})}{Q_{\text{flow}}} \qquad °\text{F}$$

TABLE 5.1   Values of C

Medium	C (Btu/lb-°F)
Air	
C_p	0.24
C_v	0.17
Water	1.00
Steel	0.12
Oil	0.50

For example, a 4000-hp motor has 40 hp or $40(0.746) = 30\,\text{kW}$ dissipated as heat. What is the temperature rise of the air if the flow through the motor is $3000\,\text{ft}^3/\text{min}$?

$$\Delta T = \frac{3715(30)}{3000}$$
$$= 37.2\,°\text{F}$$

Another form is the temperature rise of an oil reservoir with the heat dissipated by the sides:

$$\Delta T = \frac{1340(\text{kW})}{A} \quad °\text{F}$$

Here ΔT is the temperature rise between the oil and ambient air (°F), kW represents the power dissipated into the oil, and A is the surface area of the tank sides (ft²).

5.3 RADIATION

Thermal radiation is electromagnetic radiation emitted by a body as a result of its temperature. Radiation heat transfer includes the influence of material properties and the geometric arrangement of the bodies on how the energy is exchanged. Basically, the geometric factor or shade factor involves determining the portion of energy that leaves one surface and reaches the other. In this book we do away with all the complications by using a simplified radiation equation. Unfortunately, if its limitations are not understood, it also does away with much of the accuracy of the results. It is based on the absorbing body being small relative to its surroundings, much like a steam pipe radiating to a room.

$$q_r = \sigma \varepsilon_S A_S (T_S^4 - T_L^4) \quad \text{Btu/hr}$$

Here T_S, ε_S, and A_S (ft²) represent the smaller body, and T_L is the temperature of the surroundings. The temperatures in the radiation equation are absolute temperatures (i.e., add 460° to the Fahrenheit scale), and σ is a constant:

$$\sigma = 1.74 \times 10^{-9} \text{ Btu/ft}^2\text{-hr-°R}^4$$

ε_S is the emissivity and is a tabulated value for different materials and has a range of zero to 1, with values shown in Table 5.2. In many engineering problems the temperatures usually need to be high for radiation to become predominant over conduction and convection, and should be checked, especially when temperatures are above 500°F or 960°R.

TABLE 5.2   Emissivity Values

Material	ε_S
Glass	0.94
Lampblack	0.80
Polished aluminum	0.04
Mild steel	0.20

5.4   HEAT SOURCES

We have talked of several modes of dissipating heat, such as conduction, convection, and radiation. Now sources of heat input will be reviewed along with conversions, so that Btu/hr can be calculated.

Electrical and Mechanical Losses   Nothing is 100% efficient, and the efficiencies of various devices are usually tabulated. Inefficiencies of machines are usually given off as heat, and this heat has to be dissipated by convection or radiation. Consider a motor of 3000 hp that is 98% efficient. If all inefficiency were due to resistance and resulted in heat, the horsepower loss would be

$$loss = 3000(0.02) = 60 \, hp$$

Since there are 2544 Btu/hr per horsepower,

$$loss = 60(2544) = 152,640 \, Btu/hr$$

The same approach can be used for a mechanical gearbox with an efficiency of 95%. Other useful conversions follow:

watt =	$VI = I^2R$
0.746 kW =	1 hp
746 watts =	1 hp
1 Btu =	778 ft-lb
1 kW =	3410 Btu/hr
1 hp =	33,000 ft-lb/min
1 hp =	42.2 Btu/min
1 ton of refrigeration =	12,000 Btu/hr

Much that we have learned can now be used. For example, if it were desired to know what loss might be expected from a shaft in a sleeve bearing, the following approach could be used to build a model:

$$hp = \frac{T(rpm)}{63,000}$$

Given torque T (in.-lb), shaft diameter D (in.), weight of shaft on bearing W (lb), friction coefficient μ, and rpm the shaft speed, we have

$$T = \frac{FD}{2}$$
$$F = \mu W$$

Putting this all together, the loss that needs to be removed by some heat transfer mode is

$$\text{hp} = \frac{\mu W (\text{rpm})(D/2)}{63,000} \qquad \text{horsepower loss}$$

5.5   CASE HISTORY: INSULATION BURNOUT OF A RESISTOR BANK

This case history relates to a bank of resistors that was used in a large motorized vehicle and was mass produced. The problem was that the wires kept burning off, and a better understanding of the system was required to initiate a fix. Warranty claims were expected and an inexpensive fix was being reviewed. The following model was developed to help understand the problem.

The resistor bank is shown in Figure 5.2, and the bank is enclosed front and back, with a plate to protect them from damage. To simplify the problem, assume that each resistor is far enough from the next that it isn't influenced by it. Also neglect conduction through the bus, as the area was small. The heat transfer only needs to be considered in one resistor and is shown shaded within the dashed box.

Energy balance:

$$q_{\text{generated}} = q_{\text{conduction}} = q_{\text{convection}} + q_{\text{radiation}}$$

$$\frac{I^2 R \text{ (watts)}/746 \text{ watts}}{\text{hp}} \times \frac{2544 \text{ Btu/hr}}{\text{hp}} = q_{\text{convection}} + q_{\text{radiation}}$$

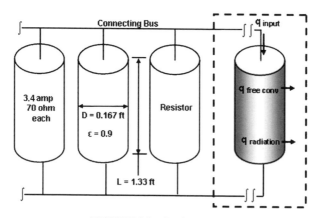

FIGURE 5.2   Resistor bank.

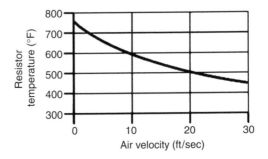

FIGURE 5.3   Resistor temperature.

$$2{,}760 = (1 + 0.225\ V)A\,\Delta T_{\text{resistor}} + 0.16 \times 10^{-8} A(\Delta T_{\text{Rankin}})^4$$

$$A = \pi DL \qquad \text{ft}^2$$

Selecting values for $\Delta T_{\text{resistor}}$ and using trial-and-error techniques yields, with ambient 100°F and no air flow;

$$\Delta T_{\text{resistor}} = 750°F$$

$$q_{\text{free convection}} = 454\ \text{Btu/hr}$$

$$q_{\text{radiation}} = 2{,}280\ \text{Btu/hr}$$

This would be enough to cause insulation problems. Extensive testing was conducted and indicated an in-service temperature of the resistor of 840°F.

Field insulation was added at selective locations along with some flow of air over the resistor. The effect of airflow over the resistor is shown in Figure 5.3. Some things learned from this analysis:

1. A fan should help, since increasing the velocity is significant.
2. The simple model gave one a good understanding of the problem. It indicated that the expense of experimental testing on an actual piece of equipment was warranted, even at a cost of $10,000.
3. Heat transfer is not an exact science.

5.6   CASE HISTORY: EMBEDDED BEARING TEMPERATURE

Here's another actual example on using energy balances. A new bearing design was proposed to be mounted in a plastic cover instead of an aluminum cover (Figure 5.4).

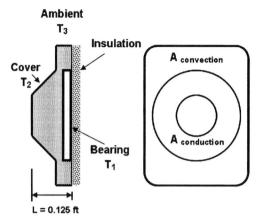

FIGURE 5.4 Embedded bearing.

The cover temperature, T_2, was measured as 130°F, and the outside temperature was 80°F. Only heat flow toward the outside was possible. Whether the plastic would insulate the bearing and not dissipate enough heat was a concern, so some idea of the bearing temperature T_1 was desired. The convection area A_{conv} was three times the conduction area A_{cond}, and the convection heat transfer coefficient h in still air was 1.5. Units are ft, hr, °F, and Btu.

Calculating the plastic cover bearing temperature by an energy balance:

$$\text{heat conducted through } L = \text{heat convected from surface}$$

$$\frac{k}{L} A_{cond} (T_1 - T_2) = hA_{conv}(T_2 - T_3)$$

$$T_1 - T_2 = h \frac{L}{k} \frac{A_{conv}}{A_{cond}} (T_2 - T_3)$$

$$= 1.5 \left(\frac{0.125}{0.3} \right)(3)(130 - 80)$$

$$= 94°F$$

$T_1 = 94 + 130 = 224°F$ is the bearing temperature.

Do you have any idea what the bearing temperature might be with an aluminum cover? As a first guess, you could make the assumption that because of the high k value for aluminum, T_1 is about equal to T_2. Since the heat generated by the bearing remains the same, the bearing temperature would be about 130°F, which is considerable cooler than 224°F.

How much heat is generated by the bearing if $A_{conv} = 0.3\,\text{ft}^2$?

$$q_{generated} = 1.5(0.3)(130 - 80) = 22.5\,\text{Btu/hr}$$

which isn't much.

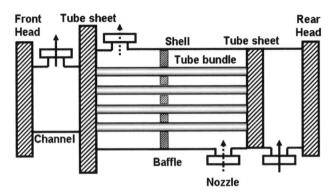

FIGURE 5.5   Fixed-tube heat exchanger.

5.7   TYPES OF HEAT EXCHANGERS

This book is not about designing heat exchangers, but deals more with troubleshooting them. To troubleshoot we have learned that it is very important to understand what goes into the design of a piece of equipment. With this knowledge we can talk intelligently with equipment manufacturers' engineering and field service personnel. Imagine trying to troubleshoot a gear unit problem if you didn't understand gearing. There are many heat exchanger designs, and three generic designs are shown because they are common in the petrochemical industry. Each design has its purpose and each design has advantages and limitations.

There may be several hundred tubes in the tube bundles shown in Figure 5.5, and the tube sheets are clamped between flanges, with gaskets providing the sealing.

Fixed-tube Heat Exchanger   The fixed tube is used extensively. It is a simple design that permits easy cleaning of the tubes. The head is removed and the inside of the tubes cleaned. One drawback of this type of exchanger is that the temperature differential between the fluids on the tube side and on the shell side is limited. The shell and tube length is fixed and will not permit thermal growth. Usually, a temperature difference of less than 200°F is desirable. Excessive temperature differential can cause the tubes to grow more or less than the shell. This can cause tubes to fail at the head connections and leak, or cause gasket leaks or ruptured tubes.

Floating-Head Heat Exchanger   This type of exchanger (Figure 5.6) is used when high-temperature differentials are required between the fluids or gases. Since the head end is free to "float" (i.e., grow), it will not cause the problems mentioned in connection with the fixed-tube exchanger. These exchangers are usually more expensive and complex than fixed-tube exchangers. The internal gaskets can eventually leak.

U-Tube Heat Exchanger   Large temperature differentials can be tolerated without the need for a floating head. The tube bundle can consist of thousands of tube loops. As the bends get smaller toward the center of the bundle (Figure 5.7), they can be

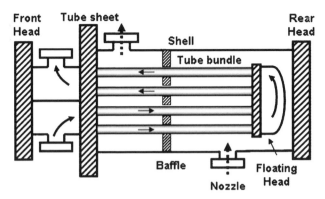

FIGURE 5.6 Floating-head heat exchanger.

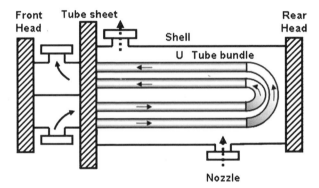

FIGURE 5.7 U-tube heat exchanger.

damaged or rupture. They are impossible to remove without removing the outer U-bends. The inner sharp turns can lead to erosion problems, so these types of exchangers typically are used with clean fluids.

5.8 HEAT EXCHANGER DESIGN

A common heat exchanger is one where a liquid is cooled or heated by water or another liquid. Several basic equations are used to size liquid heat exchangers. These relationships are discussed in this section. The temperature drop or increase in the process stream is usually a known design parameter, as is the mass flow. The desired outcome from the calculations is the surface area of the tubes required to achieve this temperature increase or decrease. Variations in use of the equations can occur, such as when an exchanger is already in place and the process conditions are being changed. In this case the area is known but the process outlet temperature is not. In either case the same heat exchanger equations would be used.

One important parameter that must be known is the overall heat transfer coefficient U. This coefficient is the combination of several heat transfer resistances and can be expressed as

$$\frac{1}{U} = \frac{1}{h_o} + \frac{t}{k} + \frac{1}{h_i(A_o/A_i)}$$

This subscription o refers to the outside of the tube or the process side fluid, and i to the inside surface. The overall heat transfer resistance ($1/U$) is the sum of the individual resistances due to the outside film coefficient on the process fluid side of the tube ($1/h_o$), the wall material thickness (t/k), and the inside film coefficient ($1/h_i$), corrected for areas. The film coefficients represent the conductivity of a thin layer of fluid that adheres to the heat transfer surface.

The difficulty in measuring individual heat transfer coefficients is one major advantage in using the overall coefficient U. It is a single parameter that is measured easily by in-service data. Typical values for U are presented in this section. Values for film coefficients (h) are needed when the actual tube temperature is required for stress calculation. The equation above allows one to back-calculate h_o if h_i is assumed very large (i.e., offers no resistance) and the areas, U, the tube thickness (t), and the thermal conductivity of the tube material (k) are known. Units are ft, lb, °F, hr, and Btu.

The following equations represent the basic heat exchanger design equations.

1. Specific heat equation:

$$q = mC\,\Delta T \qquad \text{Btu/min}$$

where m is in lb/min, C is in Btu/lb-°F, and ΔT is the temperature differential (°F).

2. Logarithmic mean temperature difference (LMTD), T_m:

$$T_m = \frac{A}{B}$$

$$A = (T_{pout} - T_{wout}) - (T_{pin} - T_{win})$$

$$B = \ln \frac{T_{pout} - T_{wout}}{T_{pin} - T_{win}}$$

The LMTD equation was developed using energy balances throughout the exchanger passes. T_m is a suitable mean temperature between the inlets and outlets of the process side and water or other liquid or the gas side and is used later; ln is the logarithm to the base e.

3. Calculate the correction factor for one shell pass and multiple tube passes:

$$R = \frac{T_{win} - T_{wout}}{T_{pout} - T_{pin}}$$

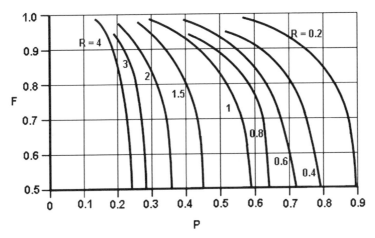

FIGURE 5.8   Correction factor F chart.

$$P = \frac{T_{pout} - T_{pin}}{T_{win} - T_{pin}}$$

Find factor F from Figure 5.8.

4. Convection heat transferred:

$$q = \frac{UAFT_m}{60} \qquad \text{Btu/min}$$

where U is the overall heat transfer coefficient (Btu/hr-ft²-°F) and A is the tube surface area (ft²).

Heated or cooled	U
Cool gas with water	20
Cool hydrocarbons with water	60
Heat hydrocarbons with steam	100

5. Tube surface area A:

$$A = \frac{\pi N_{tp} D L N_p}{144} \qquad \text{ft}^2$$

where N_{tp} is the number of tubes per pass, D the tube diameter (in.), L the tube length per pass (in.), and N_p the number of passes.

Use of these equations is best demonstrated by a case history.

5.9   CASE HISTORY: VERIFYING THE SIZE OF AN OIL COOLER

This system is interesting, as it is fairly typical and also illustrates the use of energy balances. In this example the small lubrication system shown in Figure 5.9 is proposed to provide lubrication to the bearings of a 2000-hp steam turbine. The intent is to eliminate the internal lubrication system, which has proven unreliable. On reviewing the system obtained from the equipment supplier, the proposed oil cooler appeared smaller than what the engineering group felt was necessary. The tube surface area of the exchanger selected is known. The problem becomes one of using the design equations to determine the tube surface area required and to compare it with the design proposed.

Usually, the heat load will have to be defined, but in the case being analyzed, this is available since the inlet and outlet temperatures of the oil, along with the mass flow, is known. Also known is the mass flow of water and the inlet temperature of the water. This is all the information required to determine the surface area required for the exchanger.

A useful equation relates flow in gal/min to lb/min:

$$m = 8.3(\text{specific gravity})(\text{gal/min}) \qquad \text{lb/min}$$

The numbering corresponds to that of the list in Section 5.8.

1. Heat transferred with oil having a C of 0.46:

$$m = 8.3(0.88)(10) = 73 \text{ lb/min}$$
$$q = mC\,\Delta T$$
$$= 73(0.46)(150 - 120)$$
$$= 1007 \text{ Btu/min}$$

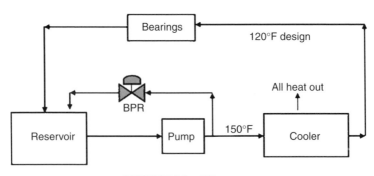

FIGURE 5.9   Oil system.

This is the total heat that will be removed by the water. To find T_{wout}, use the equation again with the q above. The water flow is 20 gal/min and $T_{win} = 100°F$.

$$m = 8.3(1.0)(20)$$

$$= 166 \text{ lb/min}$$

$$q = mC\Delta T$$

$$1007 = 166(1.0)(T_{wout} - 100)$$

$$T_{wout} = 106°F$$

2. Calculating the LMTD:

$$T_m = \frac{A}{B}$$

$$A = (T_{pout} - T_{wout}) - (T_{pin} - T_{win})$$

$$= (120 - 106) - (150 - 100)$$

$$= -36$$

$$B = \ln \frac{T_{pout} - T_{wout}}{T_{pin} - T_{win}}$$

$$= \ln \frac{120 - 106}{150 - 100}$$

$$= -1.273$$

$$T_m = \frac{-36}{-1.273}$$

$$= 28.2$$

3. Calculating correction factor F:

$$R = \frac{T_{win} - T_{wout}}{T_{pout} - T_{pin}}$$

$$= \frac{100 - 106}{120 - 150}$$

$$= 0.2$$

$$P = \frac{T_{pout} - T_{pin}}{T_{win} - T_{pin}}$$

$$= \frac{120 - 150}{100 - 150}$$

$$= 0.6$$

$$F = 0.97 \qquad \text{from the graph}$$

4. Using an overall heat transfer coefficient $U = 60$ yields

$$q = \frac{UAFT_m}{60}$$

$$1007 = \frac{60A(0.97)(28.2)}{60}$$

$$A = 36.8 \ ft^2$$

This is the surface area of the exchanger required to meet design conditions.

5. The unit supplied has $60\text{-}\frac{1}{2}$-in.-diameter tubes 24 in. long and is a single-pass exchanger:

$$A = \frac{\pi N_{tp} DLN_p}{144}$$

$$= \frac{\pi(60)(0.5)(24)(1)}{144}$$

$$= 15.7 \ ft^2$$

This is less than half of the surface area required to meet the design conditions, and a larger exchanger will have to be selected.

5.10 CASE HISTORY: TEMPERATURE DISTRIBUTION ALONG A FLARE LINE

A length of flare pipe 2200 ft long was used to connect several plant vessels and to depressurize the vessels to the flare in the rare case of emergencies. The product into the line was at $-50°F$. To eliminate a brittle fracture concern, the first 1000 ft of the piping was made of stainless steel (SS), since it had adequate toughness for this low temperature. The last 1200 ft was made of carbon steel (CS), which was adequate for $-20°F$ and above. This was done for cost-saving purposes. Process changes would lower the inlet temperature to $-73°F$ and a question was raised as to whether the piping was still acceptable for the conditions proposed.

    This case history is presented to show the use of a temperature distribution equation, which can be helpful in troubleshooting, as it was here. It is desired to know if the gas temperature reaches $-20°F$ before it is in contact with the carbon steel piping. For a length of pipe with a temperature of the gases or fluid at the start, T_{in}, and with the ambient temperature, T_{amb}, the temperature at any point along its length is $T(x)$.

$$T(x) = T_{amb} + (T_{in} - T_{amb})e^{-\varphi(x)}$$

where

$$\varphi = \frac{h\pi D}{mC}$$

Here

$$T_{amb} = 43°F$$
$$T_{in} = -73°F$$
$$h = 1.5 \text{ Btu/(hr-ft}^2\text{-°F)}$$
$$D = 5.5 \text{ ft}$$
$$m = 220,000 \text{ lb/hr}$$
$$C = 0.32 \text{ Btu/lb-°F}$$

From Figure 5.10 it appears that an inlet temperature of $-70°F$, which is shown as the solid curve, will be colder than $-20°F$, even in the carbon steel pipe. The original inlet temperature of $-50°F$, represented by the dashed curve, would have met the carbon steel limit, with the assumptions made. Mixing with warmer gases along the flare line was necessary so that the carbon steel piping temperature limits would not be violated.

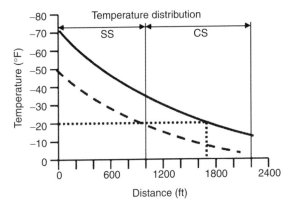

FIGURE 5.10 Temperature distribution flare line.

5.11   CASE HISTORY: DERIVATION OF PIPE TEMPERATURE DISTRIBUTION

Derivations of this complexity are not necessary in this book, as the final equations are provided. This one is being included to show how the flare-line example was solved and also to help explain the logarithmic mean temperature difference. Reviewing Chapter 5 will help in understanding this derivation.

Using Figure 5.11, the heat absorbed by the cold fluid equals the heat given up by the hot fluid. From

$$dq = mC\,dT$$

we have

$$m_c C_c\,dT_c = m_h C_h\,dT_h$$

The heat added to the cold fluid results in a temperature rise and is equated to the heat transferred through the increment of surface area of length dx:

$$m_c C_c\,dT_c = h(A)\Delta T\,dx$$

$$dT_c = \frac{hA}{m_c C_c}\,\Delta T\,dx$$

$$0 = dT_c - \frac{hA}{m_c C_c}\,\Delta T\,dx$$

Since we are talking of the same ΔT and it is just the heat lost from the hot fluid, we have

$$m_h C_h dT_h = h(A)\Delta T\,dx$$

$$0 = dT_h - \frac{hA}{m_h C_h}\,\Delta T\,dx$$

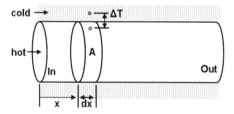

FIGURE 5.11   Heat transfer model.

Equating the equations that equal zero, we find that

$$dT_c - \frac{hA}{m_c C_c} \Delta T \, dx = dT_h - \frac{hA}{m_h C_h} \Delta T \, dx$$

Recognizing that

$$dT_h - dT_c = d(dT_h - dT_c) = d(T_h - T_c) = d \Delta T$$

and making the substitutions yields

$$\frac{d\Delta T}{\Delta T} = -h(A)\left(\frac{1}{m_h C_h} + \frac{1}{m_c C_c}\right) dx$$

For our case the heat of the cold fluid is just a fixed ambient temperature, and the solution of the differential equation can be solved by integrating both sides:

$$\int \frac{d\Delta T}{\Delta T} = \int -\frac{hA}{m_h C_h} \, dx$$

Integrating the temperature from T_{in} to T_{out} and x from $x = 0$ to x, letting $A = \pi D$, and recalling from tables of integrals that

$$\int \frac{d\Delta T}{\Delta T} = \ln \frac{\Delta T}{\Delta T}$$

$$\ln \frac{T_{out} - T_{amb}}{T_{in} - T_{amb}} = -\frac{h \pi D}{m_h C_h} x$$

The log term ln is similar to the logarithmic mean temperature difference (LMTD) that is used in Section 5.8 for heat exchangers.

In our case, interest is in the temperature distribution along the length, so an identity can be used:

When $\ln a = b$, the following is also true $a = e^b$:

Making this substitution, the final temperature distribution along the pipe is

$$T_{out\,x} = T_{amb} + (T_{in} - T_{amb})e^{-(h\pi D/mC)x}$$

This equation produces the temperature distribution along a pipe. A fluid is flowing through a pipe of diameter D with a weight flow of m and a specific heat of C. The heat is being transferred to the surrounding ambient temperature T_{amb} through a heat transfer coefficient h and its inlet temperature is T_{in}. This is the equation used in the preceding case history.

One of the advantages of a closed-form solution such as this is that it is quite simple to perform a sensitivity analysis. For example, what will happen to the distribution on hot or colder days, at different flow rates, or at different inlet temperatures?

6

COMPRESSOR SYSTEMS AND THERMODYNAMICS

Which discipline one is working with defines which equations are used most. In machinery, especially compressors, where gas compression, flows, and refrigeration are of interest, the gas laws and energy equations find wide use.

6.1 IDEAL GAS LAWS

At constant temperature:

$$\frac{P_1}{P_2} = \frac{V_2}{V_1}$$

With constant temperature the volume of an ideal gas varies inversely with the absolute pressure (psig + 14.7) = psia.

At constant pressure:

$$\frac{T_2}{T_1} = \frac{V_2}{V_1}$$

With constant pressure the volume of an ideal gas varies directly, as the absolute temperature (°F + 460) = absolute Rankine (°R).

Analytical Troubleshooting of Process Machinery and Pressure Vessels: Including Real-World Case Studies, by Anthony Sofronas
Copyright © 2006 John Wiley & Sons, Inc.

At constant volume:

$$\frac{P_2}{P_1} = \frac{T_2}{T_1}$$

With constant volume the absolute pressure of an ideal gas varies directly with the absolute temperature.

Combining the equations above produces the *equation of state*, which is easy to remember, and when one set of the variables is constant, they cancel out. This equation then results in one of the previous ones.

$$\frac{P_1 V_1}{T_1} = \frac{P_2 V_2}{T_2}$$

When dealing with the compression of gas in the *adiabatic process* (i.e., no heat is added or removed during the compression process), the following equation is used:

$$P_1 V_1^k = P_2 V_2^k$$

Here k is equal to the specific heat ratio C_p/C_v, and k values for several frequently used gases are given in Table 6.1. Using some algebra and the previous relationships, a useful form of the equation is

$$\frac{T_2}{T_1} = \left(\frac{P_2}{P_1}\right)^{(k-1)/k}$$

This produces the temperature ratio as a function of the pressure ratio. Thus, knowing the suction temperature T_1 of a compressor and the pressure ratio, the discharge temperature T_2 can be calculated. In actuality, adiabatic compression cannot be obtained, so k is replaced by n, which is now determined experimentally. This process is called *polytropic* and considers efficiencies.

Here's an example of the use of these equations. A refrigeration compressor for a reactor is pumping ethylene and it is suspected that the discharge valves may be

TABLE 6.1   Selected k Values

Gas	k Value
Air	1.40
Propane	1.13
Ammonia	1.30
Ethylene	1.24
Nitrogen	1.40

running hot. Determine the discharge temperature from the operating conditions as follows:

$$\text{Suction temperature } T_1 = 70°F + 460 = 530°R$$

$$\text{Discharge pressure } P_2 = 358 + 14.7 = 373 \text{ psia}$$

$$\text{Suction pressure } P_1 = 88 + 14.7 = 103 \text{ psia}$$

$$T_2 = T_1 \left(\frac{P_2}{P_1}\right)^{(k-1)/k}$$

$$= 530 \left(\frac{373}{103}\right)^{(1.24-1)/1.24}$$

$$= 680°R = 220°F$$

An accurate measurement of the discharge temperature indicated that it was 248°F, not 220°F. Considering that all other values were accurate, a better value for k, now n, can be found:

$$\frac{708}{530} = \left(\frac{373}{103}\right)^{(n-1)/n}$$

$$n \approx 1.29$$

6.2 CASE HISTORY: NONRELIEVING EXPLOSION RELIEF VALVE

To show the versatility of the gas laws and that they are not only for use on compressors, here's a rather imaginative use of the gas law. Some years back, an investigation was being conducted on an explosion that occurred in a cast iron manifold. The manifold was only supposed to handle air at a few psi, but process gas in the form of ethylene unfortunately found its way into the manifold and mixed with the air. There was an ignition source and the mixture exploded, sending cast iron shrapnel in all directions. Luckily, no one was injured by the flying fragments, and only minor damage was done to the equipment. The incident investigation team wanted to know why the relief valve, which was installed for such a remote possibility, didn't function to relieve the pressure. This was a very important question, as there were 23 more identical engines with identical manifolds and relief valves at this location. An immediate answer was required or the unit would be shut down, due to the significant consequences of another failure.

The relief valve was set at 10 psig since there was a steady-state manifold pressure of 5 psig air pressure acting on it. It was thought by some of the team that it should be lowered to 7 psig. The question was: Would it relieve at this pressure should another such explosion occur? Following is the two-hour analysis that helped answer the question.

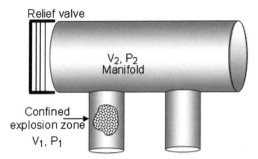

FIGURE 6.1   Manifold and relief valve system.

Consider Figure 6.1, the system being investigated. When it is set up as shown, it is fairly obvious that the gas laws are going to be helpful. The explosion occurred very rapidly, and since gas pressure and volume changes are involved, an *isentropic* process (i.e., no heat transfer) is assumed. The explosion occurred in the confined volume V_1 at a pressure of P_1. The gas then expanded into the manifold V_2 at some pressure P_2:

$$P_1 V_1^k = P_2 V_2^k$$

$$P_2 = \frac{P_1}{(V_2/V_1)^k}$$

For this problem $V_1 = 2\,\text{ft}^3$, $V_2 = 8\,\text{ft}^3$, and $k = 1.3$.

The explosion pressure of gas and dust explosions is in the range 20 to 100 psig, according to tabulated explosion data. Let's assume the worst:

$$P_1 = 100 + 15 = 115 \text{ psia}$$

$$P_2 = \frac{P_1}{(V_2/V_1)^k}$$

$$= \frac{115}{(8/2)^{1.3}} = 19 \text{ psia}$$

$$\approx 4 \text{ psig}$$

Now the pressure calculated in the manifold due to the explosion is 4 psig, which is much lower than the explosion pressure of 100 psig. This is due to the small volume expanding into a larger volume, with the resulting lower pressure. Trying to reduce the relief pressure from 10 psig to 7 psig probably won't be effective at the smaller volume explosion point.

The solution to the problem was to eliminate the potential for another explosion to develope by ensuring that ethylene could not get into the manifold. One other fix that was implemented was that the manifolds were fabricated out of steel, so that the weak link in terms of pressure was the relief valve. The moral of this analysis is that you don't have to get too analytical to solve important real-life problems in a timely manner. A detailed analysis of this problem could have taken much longer; but the final decision would have been the same.

6.3   ENERGY EQUATION

In thermodynamics it is convenient to isolate a region and call it a *system*. This was also evident in Chapter 4 when Bernoulli's equation was explained. By using the laws of conservation of energy and continuity of mass flow, the steady-state energy equation is developed. A form of the energy equation suitable for use with compressors, steam and gas turbines, and refrigeration systems is

$$\frac{Z_1}{778} + \frac{V_1^2}{2g(778)} + u_1 + \frac{P_1 v_1}{778} \pm \frac{W_{12}}{778} \pm Q_{12}$$

$$= \frac{Z_2}{778} + \frac{V_2^2}{2g(778)} + u_2 + \frac{P_2 v_2}{778}$$

The real key to the engineering use of this equation is in understanding what the terms mean and therefore knowing when certain terms can be neglected.

Due to the importance of this equation, time will be spent discussing each term in terms of the energy associated with each pound of fluid that enters the system (Table 6.2). Notice that each term of the equation has units of Btu/lb. If the equation is multiplied by 778 ft-lb/Btu, the units become ft-lb/lb, which is convenient in some cases, such as in the compressor head equation.

In the energy equation the term $u + Pv/778$ comes up often and therefore has been tabulated in tables of gas properties. It is called *enthalpy* and uses the symbol h. The energy equation takes many forms; however, it is much better that the engineer understand the equation and adapt it to the particular requirement.

TABLE 6.2   Term-by-Term Explanation (Btu/lb)

Term	Discussion
$Z/778$	Potential energy term above some datum plane for 1 lb of fluid
$V^2/[2g(778)]$	Kinetic energy term of 1 lb of fluid due to its velocity
u	Internal energy term stored in the form of thermal energy
$Pv/778$	Flow energy term of the force required to push each pound of fluid having a specific volume v (ft³/lb) into the system at a constant pressure p (lb/ft²)
$\pm W_{12}/778$	Mechanical work added or removed
$\pm Q_{12}$	Heat added (fired boiler) or removed (cooling coils)

6.4   CASE HISTORY: AIR CONDITIONER FEASIBILITY STUDY

Here's an interesting example that utilizes the specific heat equation and the energy equation in an air-conditioning feasibility study. It was desired to size the compressor for an air-conditioning system for a small aircraft. The power requirement and size were important, as the unit was to be operated off the existing electrical system, and weight was critical. This meant that the motor and compressor had to be compact. Some design constraints:

- Weigh less than 30 lb
- Cool from 130°F to 90°F in 15 minutes
- Volume to cool 60 ft^3
- 12-volt system, 10-ampere draw maximum

A preliminary test showed that the cabin temperature increased 40°F in $\frac{1}{4}$ hour. The heat influx under steady-state conditions is therefore

$$W_{air} = (60\,\text{ft}^3)(0.07\,\text{lb/ft}^3) = 4.2\,\text{lb air}$$

$$q_{air} = \frac{W_{air}}{t} c\Delta T$$

$$= \frac{4.2}{0.25}(0.24)(40)$$

$$= 161\,\text{Btu/hr}$$

Also, there will be four occupants, each generating approximately 300 Btu/hr:

$$q_{occupants} = 300(4) = 1200\,\text{Btu/hr}$$

The contents of the cabin is about 65 lb of plastic and aluminum with an average specific heat of 0.3 which also needs to have the heat absorbed:

$$q_{material} = \frac{W_{material}}{t} c\Delta T$$

$$= \frac{65}{0.25}(0.3)(40)$$

$$= 3122\,\text{Btu/hr}$$

The initial heat that has to be removed (i.e., absorbed by the evaporator) is

$$q_{removed} = q_{air} + q_{occupants} + q_{material}$$

$$= 4482\,\text{Btu/hr}$$

To determine the compressor horsepower required to accomplish the heat removal, an understanding of a simplified refrigeration system is needed. Figure 6.2

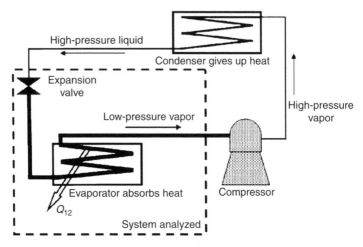

FIGURE 6.2   Refrigeration circuit.

shows the basics of such a system and is valid for small automotive, home systems, or large process refrigeration systems. In such a system two temperatures are employed: the temperature of the condenser high-pressure side that is rejected to the outside and the temperature of the low-pressure evaporator side that is associated with the cooling.

A key point is that when a liquid changes state to a vapor, it must absorb heat. In the simplified schematic this change takes place across the expansion valve and the heat in the cabin is absorbed by the vapor in the evaporator. The compressor recompresses this vapor, delivers it to the condenser where the heat is rejected to the outside, and the vapor is condensed to liquid. The cycle then starts over.

First the evaporator will be analyzed to determine the flow of refrigerant required. Reviewing the energy equation, note that

$$Z_1 = Z_2 = 0$$

$$V_1 = V_2 = 0$$

$$W_{12} = 0$$

Q_{12} = btu absorbed in evaporator per pound of refrigerant

So the energy equation reduces to

$$Q_{12} = h_{LPvapor} - h_{HPliquid} \qquad \text{Btu/lb}$$

The values for h at various temperatures and pressures in the liquid and vapor states are available from pressure–enthalpy charts [34]. Such charts show the refrigerant properties under various conditions.

At this time in the analysis, two temperatures must be selected: the liquid in the condenser and the vapor in the evaporator.

$T_{condenser}$ is set equal to the outside air temperature of 90°F, and $T_{evaporator}$ is set at 34°F and is above freezing, so that it will not freeze due to outside moisture and ice up the evaporator. From a table on refrigerants, using Freon-12, we obtain

$$h_{LPvapor} \text{ at } 34°F = 80.8 \text{ Btu/lb}$$

$$h_{HPliquid} \text{ at } 90°F = 28.7 \text{ Btu/lb}$$

$$Q_{12} = 80.8 - 28.7 = 52.1 \text{ Btu/lb}$$

The rate of refrigerant flow per 4482 Btu/hr,

$$w = \frac{q_{removed}}{Q_{12}}$$

$$= \frac{4482 \text{ Btu/hr}}{52.1 \text{ Btu/lb}}$$

$$= 86 \text{ lb/hr}$$

As more heat is removed via the evaporator, more vapor is generated which must be transferred to the condenser by the compressor. If more heat is added than can be removed by the compressor displacement, the compressor may overload.

Looking at the inlet to the compressor and the discharge, the same type of enthalpy calculations are made as were done on the evaporator. Now both conditions are vapor. From a Freon-12 table, we obtain

$$h_{suctionvapor} \text{ at } 34°F = 80.8 \text{ Btu/lb}$$

$$h_{dischargevapor} \text{ at } 114 \text{ psia} = 88 \text{ Btu/lb}$$

The theoretical horsepower to produce this flow is

$$hp = \frac{w(h_{dischargevapor} - h_{suctionvapor}) \text{Btu/lb}}{2545 \text{ Btu/hr-hp}}$$

$$= \frac{(86 \text{ lb/hr})(88-80.8) \text{ Btu/lb}}{2545 \text{ Btu/hr-hp}}$$

$$= 0.25 \text{ hp}$$

Assuming 40% losses gives us

$$hp_{shaft} = 0.35$$

The current draw is

$$\text{watts} = VI$$

$$I = \frac{(0.35\,\text{hp})(746\,\text{watts/hp})}{12\,\text{volts}}$$

$$= 22\,\text{amperes}$$

This is a pretty high current draw, and this preliminary screening doesn't look too promising. Usually, the compressor is driven off the engine, but in these types of aircraft, this was not feasible.

The analysis of this simplified system can also be used to analyze and understand larger refrigeration systems. Knowing temperatures and flows and the design process heat rejection rate, calculations can be made to determine the system's performance. Fouled evaporator coils or troubleshooting other problems can be done without tearing down the refrigeration system. A similar type of analysis can be used by considering the enthalpy in and out of steam turbines and also process gas compressors when determining the power requirements for these machines.

6.5   CENTRIFUGAL COMPRESSOR OPERATION

Centrifugal compressors are dynamic compressors that impart energy to gas molecules to compress them and move them along through the process. They are the workhorse of the petrochemical, power, and many other industries. In this book only some of the basics needed to understand these types of machines for troubleshooting purposes are reviewed.

Centrifugal compressor rotors consist of hundreds of metal parts rotating at thousands of revolutions per minute. The internal clearances are very tight, some the size of a human hair or less. Failures can cause millions of dollars in lost production costs, or lost business. This is in addition to the sizable repair costs. Therefore, the more you know about compressors, the quicker you can make corrections before something fails, or make permanent corrective fixes after it fails.

In the simple two-stage sketch shown in Figure 6.3, a molecule of gas enters the suction, into the eye of the first-stage impeller, where it is given a high velocity, exiting the impeller into the diffuser and then the return channel. It proceeds into the eye of the second-stage impeller, where the process is repeated. If there where more stages, the process would continue on.

The total pressure rise in a given stage is the sum of the dynamic pressure due to the centrifugal action of the impeller plus the static pressure due to conversion of the velocity head in the diffuser passage. About two-thirds of the pressure rise occurs in the impeller itself. The remaining increase takes place in the diffusion (i.e., velocity reduction) process. The impeller is the only means of adding energy to the gas. Diffusers, guide vanes, and return channels merely convert velocity into pressure energy. Some of the more important components are shown in Table 6.3.

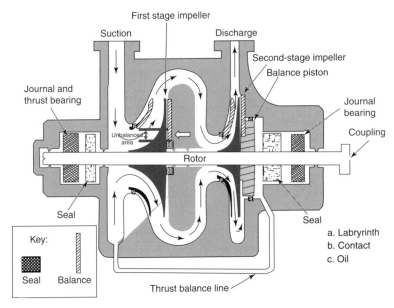

FIGURE 6.3   Simple two-stage centrifugal compressor.

One problem that can occur on compressors with balance pistons is that the balance piston seal starts to leak. This will cause the thrust balance line to get hot. A look at the sketch will reveal why. The zone behind the balance piston usually sees the suction gas that is cool, so the pipe is usually relatively cool. When the balance piston seals fail, which are shown on the outside diameter of the piston, the hot

TABLE 6.3   Selected Components and Their Purpose

Component	Purpose
Rotor and shaft	Contain rotating parts and transmit power to them
Impellers and diffusers	Impellers impart energy to the gas, and diffusers recover energy as pressure
Thrust and journal bearings	Support the rotating elements and take the weight and reaction loads
Seals	Can be gas, oil, or labyrinth (labys) seals; keep the gases and oils contained
Coupling	Provides axial and angular misalignment capabilities and connects compressor to driver
Case	Contains gases and oils and provides support for rotor
Balance piston	Relieves some of the thrust load from the thrust bearing
Thrust balance line	Provides suction pressure to the back of the balance piston

discharge gas can now flow through the line to the suction side. This causes two problems, the first being an efficiency loss due to recirculation of the discharge gas. The second can be more serious, as it will load the thrust bearing more and could result in a thrust bearing failure. This is bad, because the rotating elements can now make contact with the stationary elements and result in a wreck.

A balance piston's purpose is to provide a counteracting force to the crosshatched areas shown on the sketch. There is a force imbalance on the impellers. The balance piston is sized so that the suction pressure on the back of it will counteract most of this imbalance. This allows a smaller thrust bearing to be used. When the seal fails, all of this balance force disappears, which is why the bearing can fail. It should be mentioned that not all machines require balance pistons. Some have back-to-back impellers that cancel the forces; others have thrust bearings that can handle the unbalanced loads.

6.6   COMPRESSOR CONFIGURATIONS

One of the principal reasons that centrifugal compressors are so popular is because of their versatility. Hundreds of configurations are available, two of which are shown. This allows the process designer to fit centrifugals to almost any requirement: high pressure, low or high flow; low pressure, low or high flow and high or low temperatures. They can be used in trains of several tied together or intercooled between stages.

The straight-through flow (Figure 6.4) is a convenient arrangement that may employ as many as 12 stages. It is most often used for low-pressure process gas compression. The down nozzle configuration is shown, which is convenient, as the case can be removed without having to remove piping. The down side is that it has to be mounted on a deck so the piping beneath will clear. Up nozzles have the main inlet and outlet on the top of the case.

The double-flow configuration (Figure 6.4) doubles the possible flow capability of the compressor. The number of impellers handling each inlet flow is only half that of an equivalent straight-through machine. The down side is that the maximum head capability is also reduced. An additional benefit of the double flow is its back-to-back

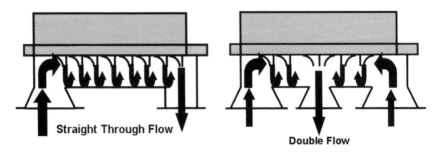

FIGURE 6.4   Straight through and double flow.

impeller arrangement, which cancels most of the thrust load. This configuration also keeps the hot discharge temperatures away from the end seals.

6.7 CENTRIFUGAL COMPRESSOR HEAD, FLOW, AND HORSEPOWER

The density of a gas is an important property, especially with gas compression. The following equation takes into account the temperature, pressure, and gas molecular weight (MW; see Table 6.4).

$$\rho = \frac{144P}{RT} \quad \text{lb/ft}^3$$

$$P = \text{psig} + 14.7$$

$$T = {}^\circ F + 460$$

$$R = \frac{1545}{MW}$$

A useful form:

$$\rho = \frac{(MW)P}{10.7T} \quad \text{lb/ ft}^3$$

When the volume flow Q (ft³/min) and the density are known, the weight flow is easily calculated:

$$w = \rho Q \quad \text{lb/min}$$

With this knowledge of thermodynamics, we have enough information to do some centrifugal compressor calculations. This isn't because we intend to size and purchase compressors, but more for understanding how it is done. Compressor performance curves are much like pump curves. Understanding them will allow much better troubleshooting results.

Compressor maps are used to define the operation of the compressor. The terms that are calculated are used to produce the map, usually by the compressor manufacturer.

TABLE 6.4 Molecular Weights

Gas	MW (lb/mol)
Air	29
Propane	44
Ammonia	17
Ethylene	28
Nitrogen	28

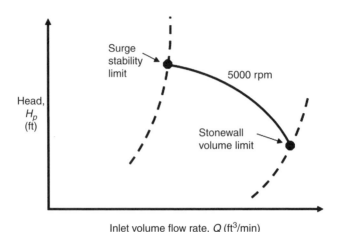

Inlet volume flow rate, Q (ft^3/min)

FIGURE 6.5   Flow–head curve centrifugal compressor.

One such flow–head curve for a fixed-speed driver system is shown in Figure 6.5. Notice that the polytropic head term H_p is in feet rather than ft-lb$_f$/lb$_m$ but means the same. It is the energy in ft-lb required to compress and transfer 1 lb of a given gas from one pressure level to another. Polytropic means that it can be restored to its original condition, or in thermodynamics jargon, it is an internally reversible process. Usually, the customer's process designer will provide the conditions that the process will require. As an example, consider that the following data are provided by the customer and the compressor needs to be selected.

$$w = 1769\,\text{lb/min} \qquad MW = 29$$
$$P_1 = 80\,\text{psia} \qquad n/(n-1) = 2.66$$
$$T_1 = 90°\text{F (550°R)} \qquad Z = 1$$
$$P_2 = 225\,\text{psia}$$

Z is called the *compressibility* of the gas and in this book is considered to be 1.

1. The inlet flow volume Q_{ICFM} (ft^3/min) is

$$\rho = \frac{(MW)P_1}{10.7T_1} \quad \text{lb/ft}^3$$

$$w = \frac{(MW)P_1}{10.7T_1} Q_{ICFM} \quad \text{lb/min}$$

$$Q_{ICFM} = \frac{w}{(MW)P_1/10.7T_1}$$

$$= \frac{1769}{(29\times80)/(10.7\times550)}$$

$$= 4487 \text{ ft}^3/\text{min at the inlet}$$

impeller arrangement, which cancels most of the thrust load. This configuration also keeps the hot discharge temperatures away from the end seals.

6.7 CENTRIFUGAL COMPRESSOR HEAD, FLOW, AND HORSEPOWER

The density of a gas is an important property, especially with gas compression. The following equation takes into account the temperature, pressure, and gas molecular weight (MW; see Table 6.4).

$$\rho = \frac{144P}{RT} \quad \text{lb/ft}^3$$

$$P = \text{psig} + 14.7$$

$$T = {}^\circ\text{F} + 460$$

$$R = \frac{1545}{\text{MW}}$$

A useful form:

$$\rho = \frac{(\text{MW})P}{10.7T} \quad \text{lb/ ft}^3$$

When the volume flow Q (ft^3/min) and the density are known, the weight flow is easily calculated:

$$w = \rho Q \quad \text{lb/min}$$

With this knowledge of thermodynamics, we have enough information to do some centrifugal compressor calculations. This isn't because we intend to size and purchase compressors, but more for understanding how it is done. Compressor performance curves are much like pump curves. Understanding them will allow much better troubleshooting results.

Compressor maps are used to define the operation of the compressor. The terms that are calculated are used to produce the map, usually by the compressor manufacturer.

TABLE 6.4   Molecular Weights

Gas	MW (lb/mol)
Air	29
Propane	44
Ammonia	17
Ethylene	28
Nitrogen	28

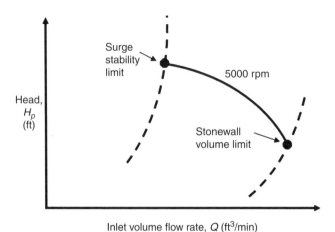

Inlet volume flow rate, Q (ft³/min)

FIGURE 6.5 Flow–head curve centrifugal compressor.

One such flow–head curve for a fixed-speed driver system is shown in Figure 6.5. Notice that the polytropic head term H_p is in feet rather than ft-lb$_f$/lb$_m$ but means the same. It is the energy in ft-lb required to compress and transfer 1 lb of a given gas from one pressure level to another. Polytropic means that it can be restored to its original condition, or in thermodynamics jargon, it is an internally reversible process. Usually, the customer's process designer will provide the conditions that the process will require. As an example, consider that the following data are provided by the customer and the compressor needs to be selected.

$$w = 1769 \text{ lb/min} \qquad \text{MW} = 29$$
$$P_1 = 80 \text{ psia} \qquad n/(n-1) = 2.66$$
$$T_1 = 90°\text{F} \ (550°\text{R}) \qquad Z = 1$$
$$P_2 = 225 \text{ psia}$$

Z is called the *compressibility* of the gas and in this book is considered to be 1.

1. The inlet flow volume Q_{ICFM} (ft³/min) is

$$\rho = \frac{(\text{MW})P_1}{10.7T_1} \qquad \text{lb/ft}^3$$

$$w = \frac{(\text{MW})P_1}{10.7T_1} Q_{\text{ICFM}} \qquad \text{lb/min}$$

$$Q_{\text{ICFM}} = \frac{w}{(\text{MW})P_1/10.7T_1}$$

$$= \frac{1769}{(29 \times 80)/(10.7 \times 550)}$$

$$= 4487 \text{ ft}^3/\text{min at the inlet}$$

ICFM represents the conditions of the gas at inlet conditions and is sometimes referred to as ACFM. In compressors this must always be used, not the flow at standard temperature and pressure, Q_{STP}.

2. The overall polytropic head H_p of the compressor is

$$H_p = \frac{1545}{MW} T_1 \frac{n}{n-1}\left[\left(\frac{P_2}{P_1}\right)^{(n-1)/n} - 1\right]$$

$$= \frac{1545}{29}(550)(2.66)\left[\left(\frac{225}{80}\right)^{0.376} - 1\right]$$

$$= 37{,}040 \text{ ft-lb}_f/\text{lb}_m = 37{,}040 \text{ ft}$$

3. Calculate the approximate gross horsepower (ghp), which doesn't include losses from leakage, seals, and so on. Also the polytropic efficiency η_p must be known, and for this example 0.76 is used:

$$\text{ghp} = \frac{wH_p}{33{,}000\eta_p}$$

$$= \frac{1769(37{,}040)}{33{,}000(0.76)}$$

$$= 2612 \text{ hp}$$

If the losses were about 5%, the driver needed at the design point would need to be at least 2743 hp.

6.8 COMPRESSOR SURGE

Surge, which is an unpleasant condition and is to be avoided on a centrifugal compressor, is shown in Figure 6.5. When the flow gets too low, it is possible for a compressor to go into surge. The importance of the variables that were just calculated will be discussed, but first let's talk a little more about surge since it can wreck compressors and can be an operator's worst dream.

In many centrifugal compressor systems, flow is controlled by a suction throttle valve, and the compressor is kept away from surge by a spill-back line. A simplified sketch of such a system is shown in Figure 6.6. As long as the suction throttling valve isn't fully open, the compressor has capacity. When the suction valve is fully open and the flow is insufficient and getting near surge, the spill-back line will dump discharge flow back to suction, cooling it first through the exchanger shown. This will increase flow and move the operating point away from surge. When the gas flow from the receiver is sufficient and the suction throttle valve is not fully open, the compressor operates away from the surge point and the spill-back line is closed.

In Figure 6.7 we describe what occurs during surge when there are no controls or if the suction throttle valve is wide open and the spill back cannot provide enough

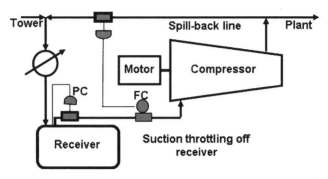

FIGURE 6.6   Spill-back line.

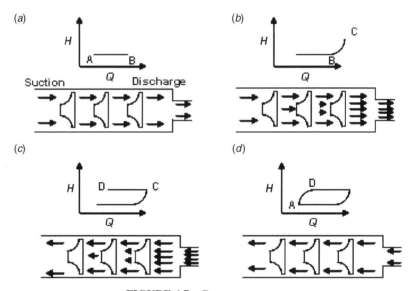

FIGURE 6.7   Compressor surge.

flow. In Figure 6.7 (*a*) normal flow through the compressor is occurring. The head is constant and the flow increases from A to B. In part (*b*) the discharge pressure begins to increase from B to C, for some reason and *H* is starting to rise. In part (*c*) the pressure is so high that the compressor cannot pump any more forward flow. An easier path for the compressor is to reverse the flow back toward the low-pressure suction end, from C to D. Doing this relieves some of the discharge pressure. In part (*d*) the head H drops from D to A at the lower flow and can start pumping toward the discharge again. This starts the entire cycle again, that is, back to (*a*). The compressor will surge like this until something is changed or the compressor thrust bearings are damaged. This will usually cause the rotating parts to contact the stationary parts, resulting in a rub or a severe wreck. These are the types of wrecks that we are called in to troubleshoot and why it is important to understand what causes surge and how to control it.

6.9 FAN LAWS

Sometimes it is necessary to estimate the performance of centrifugal compressors for operating conditions other than design. The following relationships are used to ratio the variables:

Variable	Proportional to:
Q	rpm
H	rpm^2
hp	rpm^3

6.10 FLOW–HEAD CURVE IN TROUBLESHOOTING

The fixed-speed compressor will operate along the flow–head curve for a constant receiver pressure. The suction pressure P_1 will be the value that satisfies H_p and Q_{ICFM} simultaneously on the performance chart for the compressor.

$$H_p = \frac{1545}{MW} T_1 \frac{n}{n-1} \left[\left(\frac{P_2}{P_1} \right)^{(n-1)/n} - 1 \right] \quad \text{ft}$$

$$Q_{ICFM} = \frac{w}{(MW)P_1/10.7T_1} \quad \text{ft}^3/\text{min at the inlet}$$

$$ghp = \frac{wH_p}{33,000\eta_p} \quad \text{hp}$$

As can be seen by the equations, the inlet suction pressure P_1 plays a big part on the inlet flow rate Q_{ICFM} and raising P_1 reduces Q_{ICFM}. Similarly, raising the MW, which is controlled by the suction gas composition, reduces Q_{ICFM}.

The compressor discharge pressure, P_2, is usually fixed and is not much of a factor, and since the temperatures are absolute, they have less effect. For example, a temperature increase from 80°F to 120°F changes the absolute temperature and thus the head equation by only 7%:

$$\frac{120 + 460}{80 + 460} = 1.074, \text{ or roughly 7\%}$$

MW and suction pressure changes will only change the suction throttling and spill-back flow and in this way keeps the machine from surging. Only if the compressor has no surge control or the throttle valve and spill-back flow are wide open will this be a concern.

6.11   RECIPROCATING GAS COMPRESSORS

Integral gas engine compressors have been around for a long time and they are still in heavy demand in processing plants and in the gas transmission industry. They do their jobs well. This section is not about new designs or new installations. It is about machines that have been in place for many years, have a reasonable maintenance program, and have been providing reliable service. When failures appear, they are not usually design-related problems based on the original design specifications but on something that has changed.

In this section we concern ourselves with what can go wrong to cause certain types of failures to start occurring and present brief guidelines on what can be done. The machines discussed are engine/compressors in the range 800 to 4000 hp, compressing gases up to 500 lb/in². Excluded are hypercompressors, which have unique problems and where tight tolerances and assembly alignments are critical. Electric motor–driven compressors can have similar compressor end problems but do not have the complications of the power end.

6.12   COMPONENT FAILURES AND PREVENTION

The following types of failures are discussed since they can result in costly unplanned downtimes and have been experienced by the author:

- Cracked foundations
- Cracked frames
- Galled power and compressor pistons and cylinders
- Broken valves
- Broken compressor rods and cracked compressor pistons
- Failed power-end main and rod bearings
- Power-end head gasket leaks
- Wrecked compressor end

Cracked Foundations   To understand the construction of the foundations on some of the larger systems, refer to Figure 6.8. Notice that there are three components: the engine; the block, which is usually two or three times heavier than the engine; and the mat. Usually, the block cracks in some manner rather than the mat. Figure 6.9 illustrates a search for the length of a crack by excavating along the block. By far the most troublesome is a block crack on a longer engine, since it can distort the frame and result in broken crankshafts and main and rod bearing failures. The shorter engine frames are not distorted as much.

The exact cause of a block failure is not always known, but years of startups and shutdowns with the accompanying thermal cycles, the constant pounding of the reciprocating forces, oil soaking, lack of substantial rebar in the concrete, and past wrecks certainly all play a part. There is also the case where epoxy grout has been poured on concrete that hasn't fully cured, which can create horizontal cracks at the

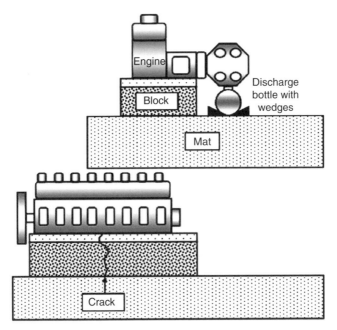

FIGURE 6.8   Engine, block with crack, and mat.

FIGURE 6.9   Excavating along a block to define a crack.

boundary. This can occur when new blocks are poured and there is time pressure to start up the unit. Usually, the epoxy is broken out and redone. In other cases it is more important to stabilize the crack than to determine the exact cause of the cracking. With an unstable crack (i.e., it shifts periodically), there is uncertainty in the amount of frame distortion that will take place. Thus, crankshaft or bearing failures cannot be predicted. Even web measurements can be meaningless, as they could change unpredictably when the engine and block are heat soaked.

When a crack cannot be seen visually, it can be verified by examining block core samples. The block can either be replaced or can be grouted using epoxy chocks which can be replaced independently as necessary or re-shimmed. Methods to stabilize cracks are with through-the-block tensioning rods, or diagonal rods in core holes filled with epoxy resin which hold the block together. These approaches need to be highly engineered by an experienced contracting firm. A faulty design could cause the block halves to misalign beyond salvage, due to the high tie-bolt loads. When oil soaked, cracks tend to become *frangible*; that is, they crumble when force is applied. With core drilling it is important to seal the core so that no oil can get within the block. It would be most unfortunate not to find any cracks and then to start some by weakening the block by drilling too many holes. This is one of the risks of core drilling that must be considered.

Cracked Frames   The primary cause of cracked frames on over 20 gas engines examined was the combination of a lightweight frame design and engine detonation, sometimes known as *power knocking*. Strain gauge measurements indicated that during detonation the frame stress in certain areas increased by over 15,000 lb/in^2 in tension. This exceeded the meehanite cast iron frame material's endurance limit. Although this design was acceptable for the original operating conditions, over the years the system had been re-rated with the addition of turbochargers but no air coolers. In the summer, with high temperatures and high loads, detonation would occur and the associated impact loading would overstress the cast iron frames and start cracks. These cracks then propagated into very large cracks. Metal stitching, a cold repair method that utilizes interlocking screws, along with through-bolt tensioning and plates, were tried with limited success on 15 engines. The final solution was to add air coolers to lower the air temperature and stop the detonation. The cracked frames were eventually converted to a heavier design. These were obtained on the used equipment market at very reasonable cost.

Galled Power and Compressor Pistons and Cylinders   These types of failures aren't unusual and the power end and compressor end have to be considered separately. At one point a rash of power-end galling was occurring (Figure 6.10). At first it was thought that it was due to the engine performance (i.e., too high exhaust temperature), but analyzing the power end showed no concern. Also, the jacket water temperature was kept below the oil temperature to maintain the desired piston–cylinder running clearance. There was concern that relined cylinders were installed, and when the manufacturer was contacted the reply was that only new solid cylinders should be used. The justification for the statement was that the sleeves could distort or cause poor heat transfer, due to the interface, and cause galling. A study was performed on historical galling failures on a large engine and is discussed in Section 7.3. It was shown that the mean time between failures of new solid cylinders and re-lined cylinders was statistically the same and didn't justify double the cost for new solid cylinders. The problem was eventually tracked down

FIGURE 6.10   Galled power-end piston.

to a new procedure being followed to re-sleeve the cylinders, which allowed warp-ing of the liner to occur. Returning to the original procedure eliminated the prob-lem on this engine.

Power end galling has also been caused by insufficient cylinder lubrication. This is easily checked with a volume check of the lubricators. Other causes have been loose internal upper block-to-frame bolts. This was determined by mounting dial indicators along the two sections and tightening all bolts. Bolts in the area of the galled piston moved 0.004 in. Galling can be a real problem, and in this case it had resulted in a crankcase explosion because of the high localized heat generated.

On the compressor end, insufficient cylinder lubrication can again be the cause, as can high temperatures breaking down the lubrication film due to leaking valves. The pumped process gas can also be a galling concern if it is a dirty service. Dilution of the lubricant and accelerated wear of the compressor end can be due to conden-sate in the cylinder. Keeping the cooling water inlet temperature regulated so that it is 15°F greater than the gas inlet temperature can help avoid this.

Broken Compressor Valves   There are many different types of valves, but depend-ing on the service they can all manage to fail at some time, and this is to be expected (Figure 6.11). They have a difficult job. However, after several years of operation, the life should be pretty well established. A series of failures that take out several valves at a time are reasons for concern and usually suggest that something has changed. Over the years numerous problems have occurred with plate-type valves

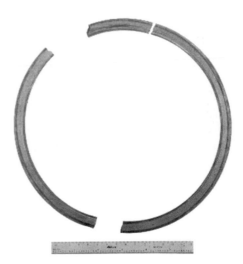

FIGURE 6.11   Failed compressor end valve plate.

with and without helical springs. The possibility of some type of pulsation-related problem due to system modifications comes to mind and indeed has been the cause at times. However, if the system has been operating in the same manner and design conditions have not changed, there is a good chance that this isn't the most probable cause. On other occasions, liquid ingestion caused valve failures, breaking the plates and springs. Draining the knockout bottles remedied those problems. Finally, there were occasions when the wrong springs were supplied to the valve rebuilder by the valve manufacturer, bench leak tests were not performed, repairs specifications were not followed, and so on. These failures were difficult to track down and caused several weeks of failures. It required having an experienced technician camp out at the rebuilder's shop. Procedures were documented, incoming material was sampled for mechanical properties, measurements tolerances and leak tests were verified, and everything was documented. This didn't make the technician the most popular person in the shop, but with his help the problem, which was basically a quality problem, was solved. The significant thing here is that the equipment owner worked with the rebuild shop rather than searching for a new one. The shop had produced a good-quality product at a reasonable cost for many years, and the equipment owner knew that they could continue doing so after the problem was addressed. The shop had serial numbers on the valves, which helped immensely when troubleshooting problem valves and tracing the valves' history.

Overtightening the cap nut holddown bolt to stop leaks cause valve failures by warping the valve seat. Loosening of the cap nut can cause extensive damage, such as failing the cast iron covers. This occurs from the valve striking the bolt, which can cause a massive leak or a damaged seating surface in the compressor cylinder housing. Care in tightening and preventive maintenance will prevent these types of failures.

Broken Compressor Rods and Cracked Compressor Pistons   These two topics are grouped together because they can have the same cause. In the course of many years of operation, one thing can be certain and that is that operations will require more throughput from the same machine. Re-rating machines by increasing the piston size or changing services to gas of heavier mole weight is common. However, the necessary calculations are not always made. One of the problems can be rod overload. The calculations are not difficult, and one such chart is shown in Figure 6.12. When these limits are exceeded, the equipment manufacturer can tell you which components can be expected to fail. For example, the higher rod loading might overload the crosshead assembly. Each engine is different, and rod overloading usually doesn't mean it is the rod that will fail.

The rod loading curve is based on the following manufacturer's equation for tension rod loading as tension was limiting:

$$\text{discharge pressure} = \frac{T_T}{A_T}$$

where

T_T = allowable tension rod load + suction pressure $\times$ cylinder area
A_T = cylinder area − rod area

The rod in compression also needs to be checked using the following equation and the curve that provides the most adverse conditions used:

$$\text{discharge pressure} = \frac{T_C}{A_C}$$

where T_C = allowable compression rod load + suction pressure $\times$ (cylinder area − rod area) and A_C is the cylinder area.

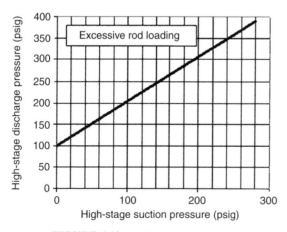

FIGURE 6.12   Rod loading diagram.

The allowable rod loading is obtained from the engine manufacturer. As an example, a condition that caused the system to operate at 220 psig suction pressure and 350 psig discharge pressure would be considered an overload. Notice that this simplified equation is just a force balance on a double-acting compressor cylinder, which considers pressure only. It is useful in a failure analysis for a check on the rod loadings. Cases have been analyzed where bearings were failing because this simple chart had been violated.

Failed pistons or rods can be attributed to the rod nut becoming loose during operation. The loosening of the piston as it reverses its stroke can result in cracks, which can eventually cause a failure at the piston–rod bore. Although the noise is usually obvious to the experienced, machines operated at a remote location and unmonitored may go undetected, with rather spectacular failures occurring. Online impact monitoring can recognize and warn of these types of problems at an early stage so that action can be taken. Although on the several problems experienced, the knocking sound has provided ample warning, it is quite possible for a failure to occur in a few revolutions. This could warrant shutdown logic to be incorporated if the risk of a wreck is unacceptable for production or safety reasons.

As with valves, liquid carryover or condensation can destroy compressor ends. Designing to API 618 should help, and if carryover is occurring, the system may need to be redesigned. Data indicate that steam tracing and insulation on the suction lines correlate strongly with a reduction in the valve failure rate, which is not surprising and should also benefit the compressor end.

Compressor nuts have a critical task and must be tightened correctly. There is not much stretch going on in the nut–rod, so any relaxation of the combination will result in a loose nut. Rolled threads and the correct nuts are a necessity, as is the proper tightening procedure. There are products on the market that are patented and use many smaller bolts to apply the correct load to the nut. These are quite useful and have been very effective as long as the manufacturer's procedures are followed carefully. Not following the correct procedure can "kink" the rod. The high precise loading required to keep the nut tight along with the confined space to work in is what makes these types of products desirable. There are times when rods don't fail in the threads and the nut doesn't come loose, as shown in Figure 6.13. This type of failure is discussed in Section 10.3.6.

Failed Power-End Main and Rod Bearings There are many types of bearings on the market, and each material composition has advantages and disadvantages. Babbitt bearings have been used for many years, since the thick babbitt layer is quite forgiving. It is relatively soft, so it can embed small hard particles without damaging the shaft and can adjust for slight misalignments. It is relatively inexpensive and can be repaired.

When the engine service changes, such as when detonation is present or excessive rod loading occurs, the use of babbitt can be a disadvantage. The pounding of the loads on the bearing can result in fatigue failures of the babbitt. It appears as spalled-out pieces of babbitt, with a thin layer still adhering to the shell backing material, which is usually bronze (Figure 6.14). When there is no thin layer remaining, this

FIGURE 6.13 Broken piston rod with nut.

FIGURE 6.14 Failed babbitt bearing.

may indicate poor bonding and is a quality problem. In the late stages it will be wiped (i.e., melted).

On higher-horsepower engines, one solution has been to use a different type of bearing, such as solid aluminum bearings. The high strength eliminates fatigue-related failures. This is a modification that has to be engineered. Since aluminum has a different thermal expansion coefficient, the crush and clearance have to be determined accurately, as does the composition of the aluminum alloy. The oil system has to be exceptionally clean since aluminum does not embed hard particles very well. Usually, prelubrication of the bearings before startup is important to develop an oil film on the steel journal and aluminum bearing.

A disadvantage is that when aluminum bearings fail, they tend to sling molten aluminum throughout the crankcase and coat the shaft with aluminum, which is quite difficult to remove. The failed aluminum bearing shows up as flakes in the oil and has the consistency of silver paint, which finds its way to the other bearings. With the correct design and monitoring, a failure will usually not get to this stage.

Power-End Head Gasket Leaks   We include power-end head gasket leaks, due to the spectacular and dangerous results they can cause. Leaks usually start out as very light "chirps" as the gas, at high velocity, escapes during the power stroke. When not tended to, the chirps turn to "barks." Every engine design has its own signature song. The tune during or after a bark can also be quite visual. In one instance it was a 3-ft flame emanating from the leak, which was quite impressive at night. Once started, the hot gas acts as a cutting torch and quickly makes rebuilding the cylinder impossible. Head gaskets in many of the older engines are made of solid copper or steel. There is not much relaxation of these gaskets when tightened correctly. The head bolts are usually quite long, and with the proper torque have adequate stretch to accommodate any embedment or thermal effects. Therefore, cyclic relaxation is not usually a problem as it is with pressured equipment such as heat exchangers in cyclic service. Usually, detonation has been the cause of leaking. This is, of course, with new gaskets and good, clean, nonwarped or cut gasket surfaces. The solution is to stop the detonation and keep the total bolt load torque higher than the firing pressure developed force, as shown in the force balance in Figure 6.15. A design margin on bolt load to a firing pressure load of 1.5 to 2.0 seems to work well for engines that experienced this problem. This is described in Section 2.17.7.

Wrecked Compressor End   Fortunately, as shown in Figure 6.16, a wreck is not a common occurrence and is discussed in Section 10.3.6. When a wreck does occur, the result is usually a pile of rubble that requires some real detective work to determine

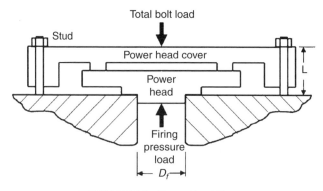

FIGURE 6.15   Power head forces.

FIGURE 6.16   Wrecked distance piece.

the cause. In this case the cause was loose slipper bolts, which eventually sheared and caused an unusual compressor rod failure. Impact monitoring probably would have caught the heavy ticking impact sound due to the cyclic reversal. By acting on the signal change, this wreck could have been avoided, along with the heavy production losses. Unfortunately, no monitoring equipment was installed. Rod failures due to loose rod nuts have been discussed and can also result in wrecked compressor ends.

Distance piece looseness to the engine frame due to bolt loosening have caused misalignment problems. Correct tightening procedures, and tightening and alignment check programs, have addressed some of these types of failures. Distance piece bolting usually consists of rather short studs or bolts with little stretch capability. Relaxation by as much as 0.002 in. can cause these studs to loosen. With loose bolts and the cyclic loads greater than the bolt load, fatigue failures in the threads can be expected.

Using higher-strength rolled thread bolts and studs can help, as can Belleville washers. All of these methods need to be well engineered. Whereas the bolts and their threads may be able to take much higher torque, seasoned cast iron frame threads cannot. Too much torque and a section of a seasoned frame may break out; similarly for Belleville washers. When well engineered, there usually is little concern, but if the highly stressed washer assembly fails (and they do), the bolt will become loose.

Problems have occurred when using wedges under compressor discharge bottles (Figure 6.8) as a method to cold level the cylinders. This is wrong, of course, because with a 48-in. distance and a discharge temperature rise of 200°F, the cylinder will be raised approximately 0.063 in. when hot (i.e., $\delta = 6.6 \times 10^{-6} \, L \Delta T$) [35]. Such severe misalignment can cause havoc on the piston and rod assembly. The wedges are simply to keep the bottles from vibrating and should only be tightened snug when hot. Following the manufacturer's leveling procedure has eliminated the heavy piston wear and rod failures occurring from this cause.

   Throughout this section the terms *engineered design, experienced technician, manufacturer's support*, and *reputable supplier* have been used. This is because they are all so critical in finding and addressing the cause of failures on integral gas engine compressors. Although we all want a high-reliability design, good operating metrics, and online monitoring for early detection and mitigation, failures still occur. When they do, it is necessary to maintain the core of personnel needed to get the plant back in operation and to ensure that the same type of failure does not happen again.

6.13   RECIPROCATING COMPRESSOR HORSEPOWER CALCULATIONS

It is often desirable to know the gas horsepower value of a reciprocating gas engine compressor to help in troubleshooting. In the following equations, the pressures are again in absolute pressure units (i.e., gauge pressure $+ 14.7 =$ psia).

Gas horsepower of a compressor cylinder:

$$\text{hp}_G = \frac{P_S Q}{229} \frac{k}{k-1} \left[\left(\frac{P_D}{P_S} \right)^{(k-1)/k} - 1 \right]$$

where for a single-acting-piston swept volume,

$$Q = \frac{\pi S}{4(1728)} [D^2 \, \eta_V (\text{rpm})] \qquad \text{ft}^3/\text{min}$$

For double-acting-piston swept volume,

$$Q = \frac{\pi S}{4(1728)} [(2D^2 - d^2) \eta_V (\text{rpm})] \qquad \text{ft}^3/\text{min}$$

$$\eta_V = 1 - C \left[\left(\frac{P_D}{P_S} \right)^{1/k} - 1 \right]$$

where

$Q = $ actual swept volume (ft^3/min)

$S = $ stroke (in.)

$D = $ cylinder bore diameter (in.)

$d = $ rod diameter (in.)

$\eta_V = $ fraction volume admitted

rpm $= $ crank revolution

$C = $ clearance volume, usually around 0.1 (10%)

$P_D = $ discharge pressure (psia)

$P_S = $ suction pressure (psia)

Using the swept volume and pressure ratio, the gas horsepower per cylinder can be evaluated. For more than one cylinder, add the gas horsepower of the cylinders.

The brake horsepower is

$$\text{bhp} = \frac{\text{gas horsepower}}{\eta_M}$$

where η_M is the mechanical efficiency: For high-speed units, $\eta_M = 0.93$ to 0.95; for low-speed units, $\eta_M = 0.95$ to 0.98. This says that you need more bhp to get the equivalent gas horsepower.

Consider a double-acting single-stage gas engine compressor: rpm $= 300$; $P_D = 373$ psia; $P_S = 103$ psia; $k = 1.29$; $S = 19$ in.; $D = 16$ in.; $d = 4$ in.; $C = 0.1$.

$$\eta_V = 1 - C\left[\left(\frac{P_D}{P_S}\right)^{1/k} - 1\right]$$

$$= 1 - 0.1 \times \left[\left(\frac{373}{103}\right)^{1/1.29} - 1\right]$$

$$= 0.83$$

$$Q = \frac{\pi S}{4(1728)}\, [(2D^2 - d^2)\, \eta_V(\text{rpm})]$$

$$= \frac{\pi(19)}{4(1728)}\, \{[(2(16^2) - 4^2](0.83)(300)\}$$

$$= 1066 \text{ ft}^3/\text{min}$$

Gas horsepower of a compressor cylinder:

$$\text{hp}_G = \frac{P_S Q}{229}\frac{k}{k-1}\left[\left(\frac{P_D}{P_S}\right)^{(k-1)/k} - 1\right]$$

$$= \frac{103 \times 1066}{229}\left(\frac{1.29}{1.29-1}\right)\left[\left(\frac{373}{103}\right)^{(1.29-1)/1.29} - 1\right]$$

$$= 715 \text{ hp}$$

Brake horsepower:

$$\text{bhp} = \frac{\text{gas horsepower}}{\eta_M}$$

$$= \frac{715}{0.97}$$

$$= 737$$

In the example the suction temperature is $T_S = 70°F$ and it is desired to know the discharge temperature T_D.

$$\text{Suction temperature } T_S = 70°F + 460 = 530°R$$

$$T_D = T_S \left(\frac{P_D}{P_S}\right)^{(k-1)/k}$$

$$= 530 \left(\frac{373}{103}\right)^{(1.29-1)/1.29}$$

$$= 708°R = 248°F$$

It is a good idea to limit the discharge temperature to below 275°F to ensure adequate packing life and to avoid lubrication oil degradation. Above 300°F, oil degradation is likely, and if oxygen is present, ignition is possible. The discharge temperature should never be allowed to exceed 350°F.

6.14 TROUBLESHOOTING RECIPROCATING COMPRESSORS USING GAS CALCULATIONS

The calculations are generally used to verify re-rates, when pressures or flows are increased. This means determining if the driver and compressor can handle the new conditions adequately or if the re-rate calculations been done properly or at all. The equations may also be used to troubleshoot. For example, if a motor is running hot or power use is high, a horsepower calculation may indicate that the discharge pressure of the system is too high. When the horsepower equation indicates that everything appears acceptable, this can also be an indicator. Maybe all that is wrong is that the motor's air filters are dirty and restrict airflow through the motor. This would not affect the horsepower but would cause the motor to run hot. Other examples that the author has experienced are shown in Table 6.5.

6.15 MECHANICAL SEALS

The mechanical seal is a sealing device that forms a running seal between rotating parts and stationary parts. Figure 6.17 shows the simplest form of this seal used in

TABLE 6.5   Troubleshooting Compressors

Symptom Noted	Cause Identified	Equation
Motor temperature and amperes high	High discharge pressure, system	BHP calculation
Low packing life	High discharge temperature, valves	Discharge temperature
High motor horsepower	Cylinder galling	BHP calculation

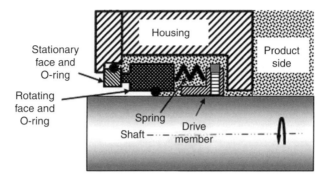

FIGURE 6.17   Single unbalanced mechanical seal.

centrifugal pumps and centrifugal compressors. It keeps liquids or gases from escaping from the casing to the atmosphere. It is a single, inside, unbalanced seal. The force of product pressure acts to push the rotating face against the stationary face for a good seal. Since it has no flush, it is limited to services handling clear fluids. Most of the pressure force is on the seal face since the areas are not balanced—thus the term *unbalanced*.

Double seals, tandem seals, and various flush arrangements are available to handle higher pressures, dirty fluids, or fluids that can solidify. A mechanical seal must seal at three points:

- A static O-ring seal between the stationary part and the housing.
- A static O-ring seal between the rotating part and the shaft.
- A dynamic seal between the rotating seal face and the stationary seal face. The dynamic seal is a thin film between the faces. This a hydrodynamic liquid film for oil seals and a aerodynamic gas film for gas seals. These faces are precision lapped for flatness of three light bands and a surface finish of 5 μin. The purpose of the spring is to keep the mating faces together during periods of shutdown or lack of product pressure.

Most seal-related problems the author has seen have been due to leaking O-rings, O-rings not free to float, poor assembly, damaged parts, or product fouling. It has been the author's experience that once the faces separate for any reason, such as vibration or misalignment, they usually will leak and have to be serviced. Seal faces and O-rings can be examined to indicate causes of failure. In Section 2.31 we discuss examining O-rings, and three common seal face causes are shown in Table 6.6.

6.16   FLEXIBLE GEAR, DIAPHRAGM, AND DISK PACK COUPLINGS

This section does not deal with coupling design but is directed at the author's experience with failure of these types of couplings. Although there are many types of

TABLE 6.6 Common Mechanical Seal Face Damage

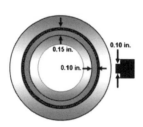

Symptom:
- Wear track wider than narrow seal face

Cause:
- Poor centering of stationary seal
- Radial runout due to bearing or shaft problem
- Shaft deflection or wobble during operation

Solution:
- Check bearing for radial runout
- Review installation techniques and instructions

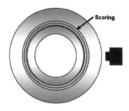

Symptom:
- Scoring or erosion

Cause:
- Rebuilding seal in dirty environment
- Faces opening/flashing/vibration or distortion
- Material found in fluid film between seal faces

Solution:
- Eliminate dry running, flashing, vibration
- Maintain clean fluid source

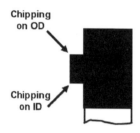

Symptom:
- Chipping outside/inside diameter

Cause:
- Faces opening, flashing operating near vapor point
- Overpressurization
- Vibration
- Product hardening and setting up

Solution:
- Ensure that product not flashing
- Proper controls and operate within parameters

couplings [36], including elastomer or quill shaft couplings for torsional tuning of systems, and rigid couplings and splined sleeve and universal joints couplings for lower-speed use, these are not discussed here. The three types considered are the most common types used in the processing industry on critical high-performance machinery: (1) gear couplings, (2) diaphragm couplings, and (3) disk pack couplings.

All couplings made by reputable manufacturers are usually trouble-free and well designed. There have only been a few instances of faulty original design, and those were quality control problems. Usually, lack of preventive maintenance, such as greasing, operational overloads, or poor installation or alignment, cause the failures. Coupling failures are often secondary, as discussed in Section 10.5.1. That failure had nothing to do with poor coupling design or poor initial alignment. Lack of keeping barrel cooling passages from fouling caused excessive misalignment due to "bowing" of the barrels.

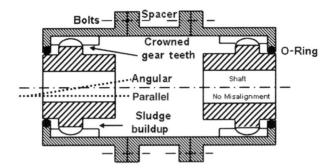

FIGURE 6.18   Gear coupling.

Gear Couplings   Gear couplings are used to connect rotating elements of two separate machine units. As shown in Figure 6.18, a typical gear coupling consists of two hubs with external gear teeth which are keyed to the ends of the driving and driven shafts. Two sleeves with mating internal teeth are bolted flange to flange. A spacer shaft is used when large misalignments are expected or when maintenance and replacement of seals is required and the equipment cannot be moved.

The coupling can be of either the *packed lube* type, which is grease packed or oil sealed, or the *continuous lube* type, in which a continuous supply of oil is sprayed on the teeth. They are similar in appearance to the sketch, but have different sealing arrangements. The coupling can tolerate combined angular and parallel offset misalignment. However, as mentioned in Section 2.18.2, caution is needed with all couplings, and going to the manufacturer's maximum can result in excessive wear and bearing loads.

These types of couplings were used extensively into the late 1970s in the petrochemical industry, but as the industry uprated to higher speeds and higher horsepower, these couplings began to fail at an alarming rate. Lack of lubrication in the greased units and the need for frequent maintenance resulted in couplings running dry and locking up; that is, without lubrication, coupling fretting and wear occurred. A zone was worn on the meshing teeth, and when thermal or other causes required an alignment adjustment, the teeth could not slide. This produces a tremendous bending moment on the shaft, which would fail the shaft in fatigue or the thrust bearings in overload.

With continuously lubricated coupling, sludging can occur. Sludging is a breakdown of the oil due to contamination of the oil. The sludge is thrown radially outward by centrifugal forces and conglomerates at the teeth. Wear proceeds and lockup can result. The results are the same as those for the grease-filled coupling.

The frictional load between gear teeth can restrict axial motion, which can load the thrust bearing. This frictional resisting load is

$$F_{\text{thrust}} = \frac{\mu T}{R \cos \varphi}$$

where

F_{thrust} = maximum thrust that can be transmitted through gear coupling (lb)

μ = breakaway friction coefficient

T = torque through coupling (in.-lb)

R = pitch of radius gear teeth (in.)

φ = pressure angle of coupling teeth, usually 20°

$$F_{\text{thrust}} = \frac{1.06\mu T}{R}$$

If $\mu = 0$ (i.e., there is no tooth friction), no thrust load could be transmitted to the thrust bearing, as the gear teeth would just slide with no resistance. Experimental work has shown breakaway friction coefficient values to be scattered over a wide range from $\mu = 0.02$ to 0.45, and they can be much higher for badly galled teeth. On average in normal practice, a value of $\mu = 0.1$ is used and would be expected with reasonable maintenance. The author has calculated this value to be very near 0.7 for badly galled teeth that resulted in shaft failure.

What this means is that with a given torque T and gear coupling size R, going from an as designed $\mu = 0.1$ to a badly galled $\mu = 0.7$ will increase the axial load that can be transmitted to the thrust bearing by a factor of 7. Thus, if a thrust bearing was designed for 2000 lb, it could now see 14,000 lb. There has to be some source for this 14,000 lb, but this means that the gear teeth won't slide until this value is reached. A locked-up coupling would be such a source.

The bending moment M_{bending} is also directly dependent on μ, θ, and gear face width (FW) [37] and could also be expected to increase, since with badly galled teeth, it is also a function of μ. This can loosen hub fits and fail shafts in fatigue, especially keyed shafts.

$$M_{\text{bending}} = T_{\text{in-lb}} \left\{ \left[\frac{0.5(\text{FW})}{R} \right]^2 + (\sin\theta_{\text{deg}} + \mu)^2 \right\}^{1/2}$$

where

M_{bending} = bending moment on shaft (in.-lb)

$T_{\text{in.-lb}}$ = torque (in.-lb)

FW = gear face width (in.)

R = coupling radius (in.)

θ_{deg} = misalignment (deg)

The coupling heat generated by friction is also affected by the coefficient of friction for a given misalignment angle θ_{rad} (radians):

$$\frac{\text{Btu/min}}{\text{tooth}} = \frac{\mu T_{\text{ft-lb}}\theta_{rad}(\text{rpm})}{183(\text{number teeth})}$$

Here $\mu = 0.1$, and values above 1 Btu/min per tooth should be reviewed in more detail.

The heat generated can increase because galling occurs. Also, increasing the misalignment, torque, or speed for a given size of coupling will increase the heat generated. If not dissipated adequately, this heat will cause the gear tooth temperature to rise. Badly galled couplings have been known to get so hot that they ignite the oil vapor in oil-lubricated geared couplings. Obviously, a high temperature will also quickly use up the grease in a greased coupling.

Consider the following application, where μ varies from 0.1 to 0.7. It shows the importance of maintaining good lubrication to a geared coupling. The loads and heat generated can increase dramatically.

$T = 31,500$ in.-lb   rpm $= 5000$
$R = 3.5$ in.   Number of teeth $= 40$
FW $= 1.75$ in. $\theta = 0.25°$ or 0.004 rad

Factor	$\mu = 0.1$	$\mu = 0.7$
F_{thrust} (lb)	950	6680
$M_{bending}$ (in.-lb)	8530	23,540
Btu/min per tooth	0.71	5.0

The limitations of geared couplings as mentioned have resulted in the use of "dry" flexible disk or diaphragm couplings on critical high-speed, high-horsepower installations. Geared couplings are still used for general-purpose installations such as centrifugal pumps, and there are many instances of them being used successfully on high-horsepower gas turbines. The appropriate engineering for such applications is a necessity, as with any high-performance coupling.

Diaphragm Couplings   In diaphragm couplings, the torque is transmitted through a single thin metal disk diaphragm, as shown in black in Figure 6.19. Angular misalignment of the coupling is accommodated by bending of the diaphragm once per revolution. Axial misalignment to accommodated shaft growth of the drive and driven machines is by oil canning or accordion-style extension or compression of the diaphragm.

A major advantage of this type of coupling is that it contains no moving parts, and no lubrication is required. There is no galling or coefficient of friction to be concerned with as with geared couplings. With proper design, the diaphragm disk and the associated welds can have stresses that will not exceed the endurance limit of the material, and thus have a theoretically infinite fatigue life.

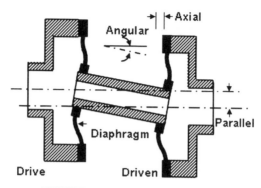

FIGURE 6.19   Diaphragm coupling.

The coupling axial stiffness is controlled by the diaphragm diameter, thickness, and profile. Axial growth of the drive and driven machines will result in axial deflection of the coupling. The axial displacements or growth of the machines must be well established and provided to the coupling manufacturer, or excessive stress could occur in the diaphragm.

These types of couplings are extremely robust, and all the failures the author has seen were bearing failures. When a bearing failed, misalignment greatly exceeding the coupling's design envelope resulted. When they do fail, diaphragm couplings usually fail due to fatigue of the diaphragm. Sometimes, in extreme overloads, the diaphragm can buckle. Diaphragms are highly engineered, are made as defect-free as possible, and are designed of materials that resist corrosion. Welds are of exceptionally high quality.

Testing of these types of couplings by manufacturers in the 1980s to determine the failure modes showed that they did fail under extreme conditions. By operating the coupling at 150% of the design torque, with axial and angular misalignment at three times that allowable, together with scratching the diaphragm, the coupling failed at 6000 rpm in five hours. Failure of the coupling was due to the diaphragm cracking at the weld.

Disk Pack Couplings   Unlike diaphragm coupling, with one spring element per side, disk pack couplings have their hubs connected to the spacer with composite springs made of thin, shimlike steel disks called *membranes*, which act as multiple leaf springs. Although they aren't designed as in Figure 6.20, the sketch provides an idea of how they are connected. The bolts, represented by the black rectangles, are usually on the same diameter. All types of misalignments are accommodated by axial deflections of the membrane sections. Any number of membranes or spring packs can be used to develop the required stiffness, and the coupling can be used with or without a spool piece.

These types of dry couplings have the same advantages as diaphragm couplings, but the multiple membranes have the added advantage that if one fails in fatigue, there are others in parallel to carry the load. The failure mode observed

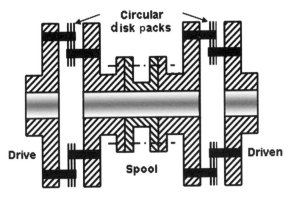

FIGURE 6.20 Disk pack coupling.

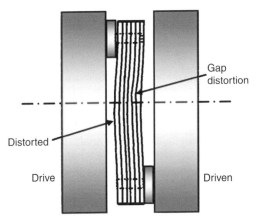

FIGURE 6.21 Distorted disk pack coupling.

has been fatigue, with some of the membrane elements found cracked. Overload causes the shim pack to buckle. Looking at the pack from the side, the observer will see a distorted or buckled shim pack (i.e., wavy in appearance) (Figure 6.21). A gap between the shim elements may also be evident. Either requires replacement of the pack. Section 8.6 illustrates the importance of the design of these disks and why only manufacturers' replacement parts should be used.

7

STATISTICS

Like the other subjects we have discussed, statistics is a big and important field. Careers are spent on small portions of this topic, areas such as sampling plans, experimental designs, and designing for reliability. All are areas of statistics that are cost-effective ways of utilizing data. Data analysis is important. In many cases it may not involve statistics at all. Knowing what data to collect, what to graph, and how to interpret results is basic to the knowledge required by all engineers and technicians.

7.1 AVERAGES, RANGE, VARIANCE, AND STANDARD DEVIATION

Probably the most used statistic is the *average*:

$$X_{avg} = \frac{X_1 + X_2 + X_3 + \cdots + X_N}{N}$$

One of the difficulties with averages is in designing with an average load. If the loads on a structure are sampled as 2000, 3000, 2000, and 10,000 lb, the average load is 4250 lb. However, the peak load is more than twice as large, and it would be prudent to design for the peak.

Analytical Troubleshooting of Process Machinery and Pressure Vessels: Including Real-World Case Studies, by Anthony Sofronas
Copyright © 2006 John Wiley & Sons, Inc.

The *range*, R_{avg}, is also a useful number, as it provides additional information on the average and its variation:

$$X_{avg} = 4250 \text{ lb}$$

$$R_{avg} = 10{,}000 - 2000 = 8000 \text{ lb}$$

Another very important measure of the dispersion of data is the *variance* (S^2):

$$S^2 = \frac{\Sigma (X - X_{avg})^2}{N - 1}$$

This calculation is done on most small, inexpensive calculators. The square root of the variance is the *standard deviation* (SD):

$$SD = (S^2)^{1/2}$$

These parameters are an indication of how much the data vary about the mean (average). The smaller the standard deviation relative to the mean, the less spread there is about the mean. This is shown in an example at the end of the section.

7.2 HISTOGRAMS AND NORMAL DISTRIBUTIONS

Tabulating data and plotting them in the form of histograms will often prove valuable. Assume that a series of measurements have been taken on rough incoming bar stock material. Eighty-one measurements were made (Table 7.1). If the data in Table 7.1 are plotted as in Figure 7.1, the bar diameters versus the number of observations forms a *histogram* or *frequency distribution*. If a smooth curve drawn

TABLE 7.1   Bar Stock Diameter and Number

Diameter Range (in.)	Number of Pieces
2.20–2.25	1
2.25–2.30	7
2.30–2.35	7
2.35–2.40	13
2.40–2.45	17
2.45–2.50	15
2.50–2.55	8
2.55–2.60	7
2.60–2.65	5
2.65–2.70	1

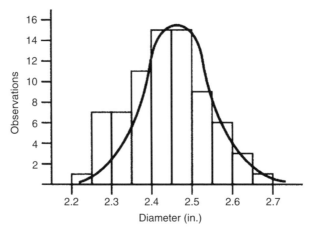

FIGURE 7.1 Normal distribution of bar stock diameter data.

through the points is symmetric and bell-shaped, the data are from a normal distribution (Figure 7.1).

Note that a quick look at data plotted in this manner provides information on:

- Estimation of the average diameter (somewhere around 2.45)
- How much the data vary about the average diameter
- The largest and smallest diameters
- What the distribution of shaft diameters look like

That the distribution is normal is important, as a large variety of engineering and manufacturing data follow a normal distribution, such as:

- Machining tolerances and size
- Clamping forces on machining lines
- Material strengths
- Valve spring hardness

The key factor in a normal distribution is that by knowing only two characteristics, the average (X_{avg}) and the standard deviation (σ) (Figure 7.2), the distribution is described completely. S is used instead of σ when a sample of the total population is being considered. The standard deviation allows a prediction on the population within an interval.

Now we can talk about the number of observations that are outside a certain σ limit (Figure 7.3):

$$X_{avg} \pm (i)\sigma$$

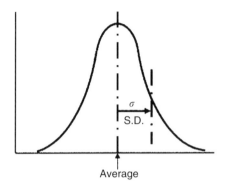

FIGURE 7.2 Average and standard deviation.

Notice that if $(i)\sigma$ were:

- 1σ, then 31 of 100 measurements would be outside $X_{avg} \pm 1\sigma$
- 2σ, then 4 to 5 of 100 measurements would be outside $X_{avg} \pm 2\sigma$
- 3σ, then 2 to 3 of 1000 measurements would be outside $X_{avg} \pm 3\sigma$

In our example of the stock material, using a calculator on the 81 (not included) data points results in $X_{avg} = 2.437$ in. and $S = \sigma = 0.100$ in., so $X_{avg} \pm 3\sigma = 2.437 \pm 0.3$. Without collecting any more data, we can say that only two or three diameters out of

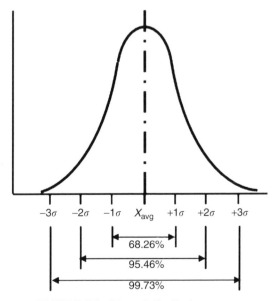

FIGURE 7.3 Normal distribution curve.

1000 will be outside the range 2.137 to 2.737. Notice how a small standard deviation relative to the mean results in tightening the range.

7.3   CASE HISTORY: POWER CYLINDER LIFE COMPARISON

An example of how averages by themselves can lead one to the wrong conclusion is illustrated by some data. The data were used in comparing the life of two types of gas engine power cylinders.

Life of solid cylinders (months):

$$16, 12, 7, 48, 14, 48, 24, 2, 24$$

Life of lined cylinders (months):

$$8, 4, 4, 5, 48, 12, 12$$

If the statistics are calculated:

Solid: $X_S = 21.7$ months of life

$$S_S = 16.5$$
$$N_S = 9$$

Lined: $X_{Lavg} = 13.3$ months of life

$$S_L = 15.7$$
$$N_L = 7$$

If only the average of the data was used, the average life of the solid cylinders was 21.7 and of the lined, 13.3 months.

One could arrive at the false conclusion that the solid cylinder achieved almost twice the life of the lined cylinder. The tip-off that the conclusion is false is the large standard deviation (S_S and S_L) relative to the mean (X_{Savg} and X_{Lavg}). To determine conclusively if X_{Savg} is truly larger than X_{Lavg}, a test on the means called *Student's t* is carried out [38]. After performing such a test, the results showed that at a 95% confidence level, X_{Savg} could be expected to be larger than X_{Lavg} only perhaps five times out of 100.

Actually, it means that you cannot say that one has a longer average life than the other. This is important, since to change a set of power cylinders costs $15,000 for new solid cylinders and $7500 for good lined cylinders. As there are eight means per overhaul, the wrong decision could cost $60,000 every four years. The outcome was to continue the test until it could be proven statistically that one was better than the other.

7.4 MEAN TIME BETWEEN FAILURES

In industry, it is important to have indices as to how equipment is operating in a plant. One useful index is the mean time between failures (MTBF) of a group of equipment or machines. Use of this index requires a good history on the failures that have occurred in the group. MTBF is defined as

$$\text{MTBF} = \frac{N}{F/M} \quad \text{months}$$

where N is the number of operating machines, F the number of failures in M months, and M number of months for total F to occur. The MTBF basically estimates the average time that one piece of equipment in the group will take to fail.

Looking at the simplest of examples, consider two machines that experience one failure in 12 months: $N = 2$, $F = 1$, and $M = 12$.

$$\text{MTBF} = \frac{2}{1/12} = 24 \text{ months}$$

So you might expect a machine to see a failure of the type defined every 24 months.

Although the MTBF is a very simple statistic to apply and contains considerable information, several precautions should be observed in its use:

1. You must be very descriptive and hold to what is defined as a failure.
2. Enough operating machines should be in the group so that one or two failures do not cause dramatic changes in the MTBF.
3. Have enough information on the MTBF so that the problems can be worked. Future MTBF trends can then be used to determine if any improvements have been made.
4. Don't try to compare MTBF data from two different plants unless the failures are defined in exactly the same manner and the services are very similar. A group of pumps pumping water can be expected to have a much higher MTBF than that of another group pumping hot hydrocarbons, even if failures are defined in exactly the same way.

The following example shows the usefulness of the MTBF method as a spot check in comparing past performance with present performance on large gas engine reciprocating compressors.

7.5 CASE HISTORY: MTBF FOR A GAS ENGINE COMPRESSOR

A gas plant contains seven compressors and is pumping a dirty gas. Several years' worth of monthly data are available on failures of all types, from loose bolts to major component failures. Several modifications have been made to the power side of the engines and several others have been proposed for the compressor side. The question asked was if the addition of air coolers to the power side during 1985 and 1986 was beneficial and whether projects proposed to clean up the gas pumped should be pursued.

Step 1: Define what will be called a failure. Defining F in the equation takes some engineering judgment. There is no use counting as failures those items that result in negligible costs or downtimes. It is also wise to include items that can indicate a possible cause of a low MTBF value. For example, in a compressor, frequent cylinder and piston replacement on the power end can be linked with detonation due to high turbocharger air temperature, overloading, or inadequate maintenance. Similarly, frequent valve failures on the compressor end could be caused by a dirty service or by faulty knockout facilities.

The following procedure for counting failures was used for this problem.

For the power end: Count as a failure if down for:

- Cylinder/piston replacement
- Cracked frame repair
- Turbocharger replacement
- Main or rod bearing replacement
- Air starter replacement
- Oil or lubricator pump replacement
- Fuel valve and head replacement if more than one

Not to be included as a failure (consecutive failures are to be counted as a single failure):

- Preventive maintenance (plugs, etc.)
- Planned overhauls
- Minor leaks or tightening of bolts
- Any repair taking less than two hours
- Troubleshooting or undetermined cause

For the compressor end: Count as a failure if down for:

- Valve replacement if more than one
- Cylinder/piston replacement
- Packing/rod replacement due to failure

Not to be included as a failure:

- Same as for the power end

Step 2: Tabulate the failure data into a useful format. In this case history, the machine was considered as two machines: the power end and the compressor end.

This is to maximize use of the failure data and because each end has unique problems. The following type of tabulation is helpful:

	MTBF (months)		
Year	Combined	Power Side	Compressor Side
1985–1986	6	9	18
1987–1988	6	18	10

Step 3: Analyze the results. Notice how when the failures are combined and called a machine failure, the MTBF is the same six months for the years being considered. If the analysis stopped at this point, the erroneous conclusion could be made that the air coolers had no effect. However, when the machine is split into power and compressor sides, different conclusions can be reached. The power side seems to have a much higher MTBF, and the compressor side a much lower MTBF. It appears that the addition of coolers has helped. The data indicated that the higher MTBF was due to fewer power cylinder and piston replacements, which would be expected with a cooler intake charge. Similarly, the compressor-end data indicated that this lower MTBF was because of more valve failures. This could also be expected with contaminated pumped gas. A proposed project to improve the gas filtering therefore appears justified.

This represents only a small part of this case history. The actual case contained 25 engine compressors in three different services, each with unique problems. The problem was simplified by division into several smaller problems in a manner similar to that done in the case just reviewed.

MTBF is extremely valuable for large groups of machinery such as pumps. Many large manufacturing companies may do monthly MTBF calculations on over 2000 pumps. With smaller groups a monthly moving average may be used to smooth out the data. This is done by calculating the MTBF on a monthly basis by using the past 12 months of data for the calculation. The next month the earliest month is discarded.

MTBF is a valuable tool but the user must heed the precautions and have a good understanding of what can be done when the MTBF value starts to shorten. A problem equipment list is one very good method for developing an improvement plan. The problem list is generated when a piece of equipment fails more than a given number of times in a year. For pumps, two failures per year are commonly used to put a pump on the list. Listed pumps will receive special attention, as improving their performance should improve the overall MTBF.

7.6 RELIABILITY

Engineers working on systems where a component failure would result in total loss of a system are usually involved with reliability [39]. Hydraulic systems for an aircraft's control surfaces and landing gear, cooling pumps for a nuclear reactor, and defense warning systems are all systems that require a very high degree of reliability.

Reliability information can also be quite useful in analyzing modifications to existing systems of machinery such as pumps.

In its simplest terms, *reliability* is the probability of a successful operation. If 10 pumps are tested and one fails, the reliability is 9 divided by 10, or 0.9:

$$\text{reliability} = \frac{\text{number of successes}}{\text{number of components tested}}$$

$$= 0.9$$

Each pump is now said to have a reliability of 0.9.

Systems consist of more than one component. When several components are in series and one of the components fails, the system will fail. This is similar to series electrical circuits or spring systems. When several components are in parallel, one failure will usually not result in total system failure. It should be obvious from the discussion above that a parallel component arrangement will have a much higher reliability than a series arrangement. The method to calculate the system reliability is illustrated next.

Arrangement of series components (Figure 7.4):

$$\text{system reliability} = R_1 R_2 \cdots R_n$$

$$= 0.9(0.9)$$

$$= 0.81$$

This is the *Law of Multiplication*, which states simply that system reliability is the product of the component reliabilities.

Arrangement of parallel components (Figure 7.5):

$$\text{system reliability} = R_1 + R_2 - R_1 R_2$$

$$= 0.9 + 0.9 - 0.9(0.9)$$

$$= 0.99$$

Notice how much more reliable the parallel system (0.99) is than the series system (0.81).

More complicated systems can be reduced to a simple series system by combining using the equations above. To illustrate combining and simplification of a complex

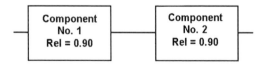

FIGURE 7.4   Series component reliability.

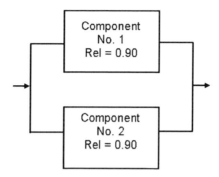

FIGURE 7.5   Parallel component reliability.

system, consider the following example, illustrated in Figure 7.6. A pumping system consists of two pumps arranged in parallel, for reliability. The additional flow gener-ated by the second pump is simply bypassed as the pump remains on as an instanta-neous backup. Each pump has a reliability of 0.9. Higher pressure is required for the system, so two booster pumps in series are proposed. Each booster pump also has a reliability of 0.9. The question is: What will happen to the system's reliability?

The overall reliability of this combination of pumps is 0.80, or stating this another way, the probability of the system working is 0.80. This is not very good. One pos-sible solution for maintaining high reliability would be to have one large booster pump with another pump in parallel. The system reliability would then be

$$\text{system reliability} = 0.99(0.99)$$

$$= 0.98$$

A reliability of 0.98 is much more acceptable than 0.80.

This example is an actual case history and was used as an argument to bolster the argument for a more desirable design. The two smaller booster pumps were being

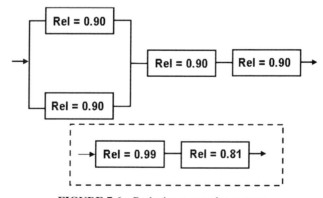

FIGURE 7.6   Reducing a complex system.

proposed because they were available and reduced the project cost by 5%. Seeing that the reliability of this important system was reduced by almost 20% was enough to get approval for the larger parallel booster pump system. Even though the management team did not have a broad background in reliability, they did know they didn't want to sacrifice the reliability of the system they now had for a 5% cost savings. They figured they would make that back if they could avoid one shutdown.

Having information such as this handy, to prove your point when in a design review meeting, is sometimes referred to as "having all your ducks in a row." It represents a very valuable outcome for the basic calculations reviewed in this book.

7.7   DETERMINISTIC AND PROBABILISTIC MODELING

The important area of deterministic and probabilistic modeling is brought up because it has been the subject of questions in seminars. Several participants who had heard of these procedures and the Monte Carlo method asked the following question: "What is the difference between deterministic modeling and probabilistic modeling?" The response given was not to explain how to do such analyses but only to explain what they are, in simple terms, using some of the basics covered in this book.

If you perform a tensile test on a round bar of diameter d and pull on it with force F, the stress can be represented as $\sigma = F/0.785d^2$ and the minimum ultimate tensile stress σ_{tensile} would be that of the bar. We would normally use the largest force F and smallest diameter d to calculate the stress and compare this with the minimum tensile strength. When $\sigma/\sigma_{\text{tensile}}$ is greater than 1, we usually say that the piece has failed. This would produce one answer and would be conservative. It is a deterministic solution.

Now suppose that instead of taking one F maximum value, the d minimum, value, and minimum σ_{tensile} measurement, we had a frequency distribution of many tests for each. From these distributions we could randomly pull samples (e.g., 100,000+) from each distribution using a technique called the Monte Carlo method and build another frequency distribution. From this new distribution we could now look at the probability of $\sigma/\sigma_{\text{tensile}}$ being greater than 1 rather than looking at a single result. In a simplistic way this is called *probabilistic modeling*. It still requires that you determine what risk you are willing to take, but you can make statements now such as that there is an 85% chance that the rod will not fail in tension. Such modeling has many uses, such as determining the reliability of complex systems based on historical data.

8

PROBLEM SOLVING AND DECISION MAKING

In problem solving, the best tool is experience. However, if the area is one where little experience is available, a calm, logical, methodical approach should be used. This will usually allow a relatively inexperienced person to solve difficult problems. Tough problems occur under situations where more than one source is contributing. Thus, there is a need for a methodical approach to isolate the important factors.

8.1 THE 80–20 RELATIONSHIP

Before suggesting a problem-solving method, I'd like to mention a rather old concept called *Pareto's law*, which states that about 80% of the total effect will come from only 20% of the members in a group. This 80–20 relationship appears with surprising regularity and is worth remembering when troubleshooting.

Consider, for example, centrifugal pump failures. Major causes include:

1. Cavitation-related problems
2. Lack of lubrication
3. Poor repairs
4. Wrong process lineup
5. Alignment
6. Defective design

Analytical Troubleshooting of Process Machinery and Pressure Vessels: Including Real-World Case Studies, by Anthony Sofronas
Copyright © 2006 John Wiley & Sons, Inc.

7. Coupling failures
8. Bad bearings/inadequate lubrication
9. Seal failure
10. Undefined

When reviewing the history of 200 pump failures at one plant, historical data indicated that causes 2, 7, and 9 were experienced most often. Not quite 80–20, but close enough (70–30) to illustrate a point. The point is to look for primary causes. The more closely you look at a problem, the more possibilities you will come up with, many of which will be trivial.

Table 8.1 represents some of the more memorable failures that the author has analyzed for the petrochemical and transportation industries. Each item consists of approximately 30 failures and certainly does not represent statistically sound sampling, but it does reenforce that many problems or failures are caused by only a few variables.

8.2   GOING THROUGH THE DATA

It is surprising how many problems are solved without use of a formal problem-solving method. Usually, parts are just replaced in kind, with the cause of failure never really

TABLE 8.1   Causes of Failures

Item	Cause of Failure (%)	Cause
Gaskets	50	Bolt load too low
	30	Improper assembly
80% of failures due to 40% of causes	10	Wrong type
	5	Bolt load too high
	5	Damaged
Gears	30	Overloading
	20	Improper lubrication
50% of failures due to 33% of causes	15	Design of gear
	15	Misalignment
	10	Improper heat treatment
	10	Poor quality control
Ball and roller bearings	35	Lack of lubrication
	20	Improper assembly
75% of failures due to 60% of causes	20	Excessive loading
	10	Defective bearing
	15	Other
Bolts and studs	35	Loose; low preload
	30	Corrosion; wrong material
85% of failures due to 60% of causes	20	Overload
	10	Wrong part design
	5	Poor quality control

TABLE 8.2   C1 Compressor History, Outages Only

Date	Equipment History
8/80	Installation and startup
9/81	Trip on low oil pressure; rag in oil pump suction
3/84	Trip on low oil pressure; bad sending unit
8/86	Trip; no oil, bad pump; spare didn't come on
2/89	System upset; unit shut down
7/89	Compressor trip on high vibration; reset
12/89	Trip on high vibration; new probes installed
1/90	Compressor wreck; bearings in bad order

understood. This is fine as long as there is never a repeat of the failure. The question is, of course: How do you know there won't be large process losses, safety concerns, or legal issues due to a similar failure in the future? Much grief can be eliminated if one would just take the time to go through the maintenance files, where the probable future operating life of a piece of equipment can be quite obvious.

For example, in the case of a small compressor failure, the shutdown data in a maintenance file were put in chronological order after a major wreck and are shown in Table 8.2. It should have been no surprise when the bearings were found to be in bad order. It seems that somewhere, someone should have examined them, but there was no mention of it in the logs. No detailed problem-solving effort was required here.

Plotting the failure data often leads to a solution, as is evident in Figure 8.1. Consider the case of an extruder screw in a barrel, with the extruder screw flight seeing all the wear. After 12 months of operation, the wear rate increases exponentially. It certainly would not be wise to operate this piece of equipment for more than 24 months since the wear could be extremely high. In some cases it was high enough to break the screw, due to lack of support. The cause was corrosion under the hard

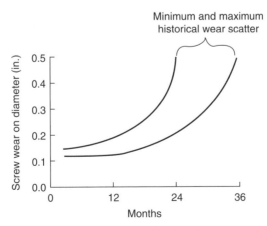

FIGURE 8.1   Extruder barrel wear.

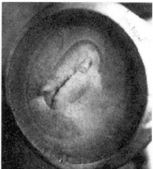

FIGURE 8.2   Extruder screw failure due to excessive wear.

surfacing which caused it to peel off. A rigid specification for applying hard surfacing solved the problem of the scatter, and the extruder wear life was extended beyond two years. Figure 8.2 illustrates a screw that failed due to a rotating bending failure caused by excessive wear.

8.3   PROBLEM-SOLVING TECHNIQUE

Before you start to solve a problem, make sure that there is one! Has something really changed, or has a baseline never been established? This isn't an empty question. Many pumps have been rebuilt at considerable expense due to a bad $10 pressure gage. Similarly, much time and money have been expended on items such as unacceptable leakage rates because no historical basis had been established.

Let's say that a problem develops. Here's a technique that has been useful in several variations:

1. Define the problem.
2. Gather data.
3. Organize and analyze the data.
4. Make an educated guess (a hypothesis) based on the data.
5. Verify your guess by tests, additional data, and calculations.
6. Implement a solution.

Actually, this is just a form of the *scientific method* (substitute *hypothesis* for *guess*).

Let's look at steps 1 through 6 in detail and then use the method in a case history.

Step 1: Define the problem. Be careful here, as many of your sources will start by defining the cause and the solution. This is okay if the problem is identical to one that has already occurred and all sources agree on the definition. Unfortunately, few problems are like that.

Step 2: Gather data.

- Gather facts and history from all sources.
- Don't assume anything.
- Screen the facts based on the reliability of the sources.
- Based on how the process, equipment, similar equipment, and so on, usually runs, determine what has changed or is different. Ask yourself if anything has changed recently. You are looking for a deviation from the norm.

Step 3: Organize and analyze the data. When the problem solvers are somewhat familiar with the equipment or process, the following table has been helpful. It is usually helpful to add any quick tests or calculation necessary to establish what the significance might be.

Observation	Significance
1. ------	• ------
2. ------	• ------

Step 4: Make an educated guess. Based on an intelligent review and screening of the data, this is where you make your best guess on what you think the cause is. Remember, it should be supported by data.

Step 5: Verify your best guess. Here is where you try to prove your guesses wrong, one by one. It's a good place to utilize some of the engineering calculations just reviewed, if applicable. One example would be a part that failed due to a cracked shaft. An overload and fatigue calculation could verify or dismiss this guess.

Step 6: Implement a solution. Solving a problem is not of much use if you don't implement a solution so that it is no longer a problem. Another way of saying this is that once the cause is identified, it must be removed.

It must be stressed here that there is usually more than one cause of a problem. The cause might be identified as an overload failure of the shaft, but what caused the overload? Was it due to inadequate procedures, poor instrumentation, or poor management of change from previous work done? All of this must be considered, and the specialist should not stop at the most obvious cause. It could reoccur and manifest itself later as a failure in some other area. Address them all! Although it may appear that only one cause was the culprit in most of the problems discussed in this book, several possibilities were addressed in search of a solution in the real world.

The best way to understand the method we are proposing is with an actual case. The six steps outlined above will be followed.

8.4   CASE HISTORY: LOSS OF A SLURRY PUMP

This example relates to a centrifugal pump that kept tripping out on overload. It was critical because production was limited when the pump was down. Unexpected tripping also caused the system to plug, which resulted in unscheduled downtime to clean. This is a very undesirable situation and one that elicits the wrath of management.

1. *Problem.* One of the four slurry pumps trip out during operation, on overload.
2. *Data* (from various sources, unscreened)
   - The same problem as in the past—it's the slurry.
   - The motor is bad; pull it out.
   - The pump is bad; pull it out.
   - The operators lined the pump up wrong.
   - There is no problem.
   - It's the process.

   A lot of other information was obtained which was not particularly useful.
3. *Organization and analysis*

Observation	Significance
• New pump and motor installed two days earlier to try to solve the problem.	• Unlikely that it is the same problem again on pump and motor.
• Not happening on other pumps.	• Same pump type, but different slurry and controls.
• Problems in the past were never really just one cause. Has been due to slurry concentration, bad motor, bad pump, air leaks, etc.	• Going to need tests to isolate this problem. This is why there are so many opinions, as there have been multiple problem causes in the past.
• Ran a quick pump curve. Found 100-ampere draw (12% above full load) at design point, on water, no slurry.	• Drawing high amperes is not a slurry problem.
• Uncoupled motor, no-load amperes 20, coasts down okay, pump turns easily by hand.	• Motor okay, pump packing not too tight, pump okay.
• Coupled motor and pump. Ran a pump curve of amperes/flow. Amperes high for flow.	• Amperes showed too high for flow indicated. Since motor and pump are okay, might be instrumentation.

4. *Educated guess.* Since the problem solvers were well versed in the system, the following was stated: "The flow is really higher than we are reading, and since

the motor and pump appear okay, it is the instrumentation." A higher flow explains the high ampere reading.

5. *Verification.* A test was run utilizing an alternative method for measuring flow. The amperes were correct for the flow calculated. The instrument was bad.

6. *Solution.* The instrument was recalibrated and put on a quarterly calibration schedule, as were similar pump systems.

Notice that the key points to solving this problem was organization of data and verification of most probable cause (guess). Also note that a pump and motor had been rebuilt and replaced unnecessarily. The cost was approximately $5000. Solving the problem took six hours.

8.5   CASE HISTORY: FATIGUED MOTOR SHAFT

A motor that is connected by a V-belt powered an agitator. The motor shaft failed in a snap ring groove, which resulted in a small fire. Data were gathered and assembled into a problem-solving chart (Table 8.3).

A well-written problem statement can help define a cause. The one shown is much better than just saying "the motor broke." Also, in problem solving it is always worthwhile to look for a deviation from normal. This is shown in steps 4 and 9.

Once the most probable cause is agreed upon, it should be tested against the observations and significance. When it doesn't hold up, you may need to reevaluate the data or develop another best guess. There were some observations, such as steps 1 and 3, that weren't really relevant. They are included since everyone's observations need to be considered. When someone is told "that's not important," it's like saying "I don't value you opinion," and it is quite unlikely that the person will contribute more information, even if it is valuable. One procedure is to review the list at the end, saying that if anyone wishes to remove his or her observation, they may. This allows irrelevant observations to be removed by the person who suggested them. The group usually welcomes this and keeps the participant participating.

The list of observations can total 50 or more, and many may not be relevant and be noted as not a factor. All of the data are recorded and checked, and new data, calculations, or testing added if necessary.

8.6   CASE HISTORY: COUPLING FAILURE

During a seminar that utilized the material in this book, a participant received a call while in class that one of the units was down, due to a failed coupling. Even though he had to leave to catch a plane back to the plant, this was an excellent time to try to troubleshoot the problem. He stayed long enough to help develop the problem-solving chart shown in Table 8.4.

It was explained that the technique used by the manufacturer of disk packs is usually proprietary information and would not be available in enough detail as a

TABLE 8.3   Problem-Solving Chart

Problem Statement:	The shaft of a motor failed in the snap ring groove.
Observation	Significance
1. Motor had been painted	Not a factor.
2. Motor and agitator turned freely	The bearings didn't fail.
3. Operator on a break	Not a factor.
4. Never failed before	Something new happened. A deviation from normal has occurred.
5. Motor shaft failed in a snap ring groove	The shaft was overstressed.
6. Metallurgy said it was a bending fatigue failure	What can cause a high bending stress in the shaft?
7. Motor belt drive can cause high load if overtightened	Belt may have been overtightened. Analyze bending stress on the shaft.
8. Bending fatigue stress okay in normal operation, but overtightening the belt could cause fatigue failure	Check when maintenance last done on belt
9. New belt installed yesterday	Enough cycles for a fatigue failure. Check how it was installed and by whom. This is a deviation.
10. Installed by untrained machinist who said he installed very tight	Adjustment screw can develop high belt loads. Probably overtightened.
Most Probable cause or Best Guess:	New machinist hadn't been trained in belt-tightening procedure. Overtightened, which caused a high bending stress and failure of the shaft.
Implement solution:	Machinists to perform only maintenance tasks for which they have been trained. Put a placard on belt guards with the correct procedure.

manufacturing specification. Designing these to withstand fatigue is how the manufacturer stays in business. Some critical specifications might be:

- Surface finish
- Surface treatment
- Material used
- Stress relieving
- Shot-peening critical areas
- Hardness used
- Hole size and edge preparation

TABLE 8.4   Problem-Solving Chart

Problem Statement: Observation	Disk pack coupling fails after installed Significance
1. Fails after installation	Not corrosion.
2. Machine hasn't just been uprated or startup procedure changed	Probably not loads or overload on startups.
3. Alignment checked and okay	Probably not alignment, but may want to hot-align check with reverse indicator.
4. Has axial play been checked?	Yes, so not this.
5. Was disk pack installed correctly?	Yes, they were made locally.
6. Have the disk pack always been made locally?	No, we used to purchase a new coupling disk pack from the manufacturer, but that was too expensive.[a]
7. Do you follow detailed specification from manufacturer to fabricate disk pack?	No, we give the local shop a sample.
8. Has the coupling failed before, and when did it start to fail?	Yes, a couple of times after a local shop started making the disk pack.[a]
Most probable cause or best guess:	Disk pack fails in fatigue due to not being designed to specifications.
Implement solution:	• Buy a disk pack from the manufacturer.
	• Perform a hot alignment check.
	• Check the failed disk for fatigue failure.

[a]Key area where deviations occurred.

- Flatness
- Heat treatment
- Final quality control

It is unlikely the local shop had this knowledge. The reason the participants came up with an answer in only 15 minutes was that the room was full of 20 knowledge-able technical people. Someone was gathering the data and organizing it into the problem-solving chart. In this case history, no calculation checks were required because and overload conditions were unlikely. The engineer who presented the problem knew the details of the history. In fairness, this wasn't his unit and he was only helping out. He went back to his plant to become a hero of sorts.

TABLE 8.5   Problem-Solving Chart

| Problem Statement: | Motorcycle won't start |
Observation	Significance
Fuel in tank?/yes	Not so simple.
Fuel on/choke on?/yes	Check ignition.
Spark at plug?/yes	Magneto and plug okay.
Any problem last ride?/no	Something new, a deviation.
Plug wet?/no	No fuel to cylinders.
Fuel in bowl?/yes	Valves or jet bad or plugged?
Educated guess:	Check for problem with valves or jet.
Implement solution:	

8.7   CASE HISTORY: MOTORCYCLE WON'T START

Not all problem solving is difficult. For example, when my son was young and I thought that he might eventually go into engineering, I tried the approach. His motorcycle wouldn't start, so we went through the method and developed Table 8.5.

The problem turned out to be a bug on the end of the fuel jet, thus allowing no fuel to the cylinder. The solution implemented was to clean the jet and add a fuel filter to the line going to the carburetor to keep bugs out. This detailed technical approach must have really impressed him, as he went into graphic design instead of engineering.

8.8   CASE HISTORY: GALLED DIE

This case is concerned with a tubelike die (Figure 8.3), which fit into a block. When the die tube was turned, it would line a certain number of holes up through which a

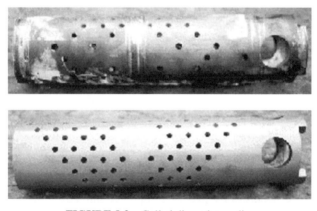

FIGURE 8.3   Galled die and new die.

TABLE 8.6   Problem-Solving Chart

| Problem Statement: | Die binds in block |
Observation:	Significance
New barrel	Not a concern.
Three new dies	Probably not dies.
New machinist	Over torque, distort? Check.
Used 1-in. impact	1000 ft-lb okay.
Same tolerance?	Check supplier.
Changed heat treatment shop	Check hardness.
Galled Rc 25, stores Rc 50	Softer.
Educated guess:	Wrong heat treatment.
Implement solution:	

polymer product would flow. The die tube was used frequently and had to turn freely. This was not the case, and the die tube would only rotate with excessive force. On removal it was found badly galled, and the actuator would not rotate it anymore. Several new dies were in the storehouse and were tried, with the same mode of failure occurring. Figure 8.3 shows a galled die and a new die, and Table 8.6 is the problem-solving chart developed.

What had happened was that the supplier had gone on strike, and during the strike another shop had done the heat treatment. Unfortunately, the shop made several with the wrong heat treat and were in Stores. See how this type of chart can cause errors. After three new dies were tried and all had galled, it was assumed that it wasn't the dies. It was the dies, and all of them were in the storehouse and all were bad. Only the tolerance checking caught the problem.

8.9   SEVEN CAUSES

The details that can be included in a formal failure analysis are evident in a book by Bloch and Geitner [40] that devotes over 600 pages to the subject. One section the author has found especially useful is "The Seven Cause Category Approach." Bloch and Geitner state that all equipment failures belong to one or more of the following seven categories:

- Faulty design
- Material defects
- Fabrication or processing errors
- Assembly or installation defects
- Off-design or unintended service conditions
- Maintenance deficiencies; neglecting procedures
- Improper operation

Reviewing the case histories we have looked at in this book should show that all causes certainly have been in one or more of these categories. As an exercise, it is recommended that the reader go through the problems in the book and identify which one or more of the seven causes were responsible. The seven causes should be reviewed on any failure to make sure that nothing was forgotten. The author has had failures where all seven causes were present.

8.10   DECISION-MAKING TECHNIQUE

In problem solving the decision comes out of the process. When there are several possible choices and your job is to pick the most appropriate, a systematic approach is required in all but the simplest cases. Any action involves risk. A good process can help reduce the risk.

In this section, a straightforward method is presented that weights and scores factors important to the decision. Since several key people can be solicited for their inputs, it becomes a team effort. Rather than going into the details of various scoring models, let's work an actual problem.

8.11   CASE HISTORY: SELECTION OF A BARREL LIFTER

In this example we select a piece of machinery to replace equipment with a poor service factor. Six pieces of lifting equipment were compared, one being the unit in use. Any unit had to cost less than $20,000, including installation. Three pieces of lifting equipment were required. For simplicity, this example covers only two cases, the cable design being the one to be replaced (Table 8.7). The factors being rated were agreed upon, as was their importance. A weight of 10 indicates that it is very important.

After the factors and weights are established, the person making the decision can review the equipment information and give it a rank. If it was excellent on satisfying the factor being considered, a 10 rating would be appropriate. Similarly, if it satisfied the factor poorly, 1 might be a good choice. The final score is simply the sum of the weight times the rating. To get this into a percent of the ideal or best that can be obtained, divide the score by the sum of each maximum weight times maximum rating and multiply by 100.

$$\text{Ideal} = 10 \times 10 + 10 \times 10 + 8 \times 10 + 8 \times 10 + 5 \times 10$$
$$+ 5 \times 10 + 10 \times 10 + 5 \times 10$$
$$= 610$$

$$\text{Cable design } 1 = \frac{239}{610}(100) = 39\%$$

$$\text{Hydraulic lift } 2 = \frac{445}{610}(100) = 73\%$$

TABLE 8.7 Decision-Making Table for Barrel Dumper Selection

Factor Rated	Weight 1 to 10	Present Cable Design	Rating 1 to 10	Hydraulic Lift	Rating 1 to 10
Operating safety	10	Binds, cable breaks	4	Restrictor for easy let down	8
Mechanical/ operating reliability	10	Above has caused poor reliability	3	Should be good simplicity	7
Ease of use	8	—	6	—	9
Low maintenance cost	8	Has been a service factor problem	2	—	6
Design simplicity	5		4	Hydraulic cylinder and chain	4
Minimum space	5	—	7	—	7
Proven design	10	New units have cable problems in 1 to 2 years	2	Much like forklift	9
Good housekeeping	5	—	6	—	6
Σweight $\times$ rating	—	—	239	—	445
% of ideal		$(239/610) \times 100$ $= 39\%$		$(445/610) \times 100$ $= 73\%$	

The other four designs done in a similar manner scored 63%, 61%, 63%, and 58%. The other designs were fair in that they were much better than the original cable design (39%).

Although your engineering judgment will control based on past experience with similar types of equipment, such a systematic organization allows you to screen the choices. As a point of interest, three of the highest scorers (73%) were purchased and have worked flawlessly with no safety incidents in 10 years. The best compliment has been that there have been no complaints from operators or machinists.

9

MATERIALS OF CONSTRUCTION

Selecting the engineering materials used to construct machines, structures, and pressure containment vessels is a highly specialized area. The effect of the environment: that is, corrosion mechanisms, temperature effects, heat treatments, welding procedures, composition, and many more factors, makes material selection a complex subject. Input from a materials engineer is usually quite welcome. What is presented here is some basic information that should be helpful to those involved in design, troubleshooting, and repair of equipment.

Temperature plays an important part in the strength of equipment. Figure 9.1 illustrates the degradation in properties as the temperature of carbon steels is raised. Components designed for room temperature and operated at elevated temperatures have been the cause of many failures. For lowering the creep rate in steels, alloying elements such as nickel, manganese, molybdenum, tungsten, vanadium, and chromium are added.

At 1000°F, the tensile strength of a medium-carbon steel is half of what it was at room temperature, and the yield is so low it would be not be usable. Our discussion of creep in Section 2.2.8 shows that to limit elongation, or in a furnace, to limit thinning of the tubes with pressure stress, stresses have to be very low at high temperatures.

One important fact to remember when materials are being selected for use in various environments is that they can all suffer from some form of degradation. For example, $1\frac{1}{2}$ and $2\frac{1}{4}$% chromium–molybdenum steels experience a loss of toughness called *temper embrittlement* when exposed to temperatures above 750°F for long periods.

Analytical Troubleshooting of Process Machinery and Pressure Vessels: Including Real-World Case Studies, by Anthony Sofronas
Copyright © 2006 John Wiley & Sons, Inc.

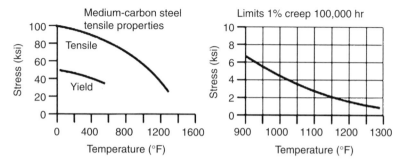

FIGURE 9.1 Effect of temperature on steels.

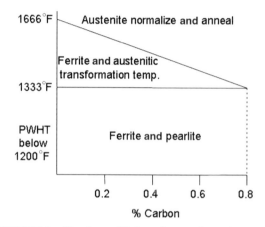

FIGURE 9.2 Simple equilibrium diagram for carbon steels.

This can result in cracks at welds or other high-stress areas. This is pointed out so that the reader will rely on a materials specialist when selecting materials for critical services.

Figure 9.2 is a very simple equilibrium diagram for steel and is presented to show the effects of temperature on carbon steels. The figure shows that the lattice structure varies from one form to another with temperature. Quenching, then cooling down too quickly can develop other structures not shown on the chart. For example, going from ferrite and austenite at 1333°F to 700°F very quickly will form a martinsitic structure, which may not be desired. Thus, temperature excursions on machines and vessels can develop unfavorable material properties, which may result in failures with the loads imposed.

Low temperatures present a new set of problems to carbon steels, and this is discussed in detail in Section 11.5. Whereas the tensile strength tends to increase some at lower temperatures, toughness as measured by the energy absorption decreases. One test that measures the energy absorbed (ft-lb) is the *Charpy V impact test*

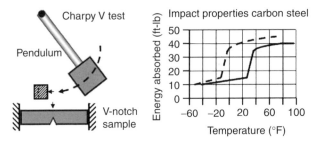

FIGURE 9.3   Charpy V impact test and graph.

(Figure 9.3). A pendulum is released from a known height, strikes a carefully prepared test specimen with a V-notch in it, passes through it, and rises to a height that is measured. From this the energy absorbed by the impact can be determined. It should be obvious that one of the problems with the test is that not all designs have notches like the test specimen. For comparison purposes it seems to be a reasonable standard.

Several other methods are also available for measuring toughness. It is interesting to note that in the 1940s during World War II, new ships using welded hull designs were breaking in half. A committee determined that the cause was brittleness of the material due to cold water, and that the welded structure allowed a crack to run. Before this time, most ships used rivet-type construction. No materials analyzed from the ships were known to have suffered brittle fracture of at least 15 ft-lb impact property at temperature. Even today, this is thought to be a good toughness value for pressure vessel steels at temperature, and has been proven by experience.

The effect on the impact properties of decreasing temperatures is quite striking. Although both steels tested (Figure 9.3) had almost the same chemical compositions, one steel had a much higher transition temperature than the other. One would be more suitable than the other for use at lower temperatures. For example, operation at 0°F (Figure 9.3) might not be a problem for one steel but would present concerns with the other steel, especially if it had notches.

9.1   CARBON STEELS

The most common carbon steels are 1020 to 1045, the last two digits representing the percent carbon (e.g., 1020 contains 0.20% C). They can be heat treated (HT) to 90 ksi tensile strength and 55 ksi yield. Weld repairs are possible without PWHT up to 0.3% C. Most steels for use in pressure vessels have a carbon content below 0.4%, as higher values tend to be brittle and difficult to weld.

A common material for welded pressure vessels is ASTM A 516, grade 70, which is carbon steel plate material (0.3 C, 0.85 manganese). The grade represents the tensile strength in kips per square inch; grade 70 has a minimum tensile strength of 70 ksi and a minimum yield of 38 ksi. It is readily weldable and has good notch toughness. Carbon steels are usually limited to temperatures below 1000°F. Various elements are added to steels to enhance their properties and some of which are indicated in Table 9.1.

TABLE 9.1   Effect of Adding Elements to Steel

Element	What It Does
Aluminum (killed steel results when silica is added to stop the reaction)	Restricts grain growth, so it improves toughness; in deoxidized steel, carbon and oxygen don't react during solidification.
Chromium	Increases corrosion resistance; adds high-temperature strength.
Nickel	Adds toughness.
Silicon	Improves oxidation resistance; increases hardenability and strength.

9.2   HIGH-STRENGTH LOW-ALLOY STEELS

These are typically chromium (up to 10%), molybdenum, and nickel-alloy steels. These elements enhance the steel for high-temperature and hydrogen service applications. 4140 (1 Cr) and 4340 (2 Ni, 1 Cr, 3 Mo) are used as shafting and gearing materials up to 450°F. ASTM A 193 B7 bolts are made from 4140 bar stock. 4140 HT typically has a tensile strength of 125 ksi and 250 BHN. It can be welded, but special preheat and temperature control precautions are required to prevent cracking. The 4340 series has nickel added to increase the toughness and fracture resistance and has much deeper hardenability than the 4100 series.

Heat treatment can make a big difference in the properties of these steels. For example, 4140, with an HT value of 800°F, can have a tensile strength of 180 ksi, and an HT to 1300°F can have a tensile strength of 110 ksi (see Section 9.10 for more on this subject). Cost per pound is about 2.5 times that of carbon steel.

9.3   MARTENSITIC STAINLESS STEELS

410 (12 Cr) and 416 have a greater corrosion resistance than that of low-alloy steel. Tensile strength/yields of 100 ksi/75 ksi, 225 BHN are typical. The material is hardenable, which can be used to reduce its tendency for galling by keeping a 50 BHN difference between contacting materials, such as when it is used for wear rings. This stainless can rust to some degree, welding can be difficult, and cost per pound is three to eight times that of carbon steel.

9.4   AUSTENITIC STAINLESS STEELS

Type 304 (18 Cr, 8 Ni) and type 316 (17 Cr, 12 Ni) are typical materials and are used for vessels and shafting. Molybdenum is added to 316 stainless steel for corrosion resistance. The low-carbon grades, 304L and 316L, are less susceptible to carbide precipitation during welding, which can result in intergranular corrosion. The yield strengths are low, typically 35 ksi, and this needs to be considered in designs. These stainless steels are nonhardenable and gall easily. Most are nonmagnetic and easily weldable. The high-nickel variety can be used down to $-320°F$. These steels are susceptible to chloride stress corrosion cracking. Generally, stress, oxygen, and a temperature above 150°F are required, and cracking is accelerated under low-pH conditions. ASTM A 193 grade B8 fasteners are 304 material and B8M is 316. Cost per pound is three to eight times that of carbon steel.

9.5 MONEL 400

The Monel 400 alloy, which contains 66% Ni and 28% Cu, is used where strength and high corrosion resistance are required; 130 ksi tensile strength, 85 ksi yield, and 250 BHN are typical values. It is used in marine shafting and processing equipment where corrosion is a problem. Its cost per pound is 17 times that of carbon steel.

9.6 17–4 PH

17-4 PH is a precipitation-hardened stainless steel containing 17% Cr and 4% Ni and is used where corrosion resistance and strength are required. Depending on the heat treatment, the tensile can range from 135 to 190 ksi. 17-4PH 1150 is air cooled at 1150°F and has a 135-ksi tensile strength; 17-4 PH 900 has 190 ksi. The 1150 series is much less susceptible than the 900 series to stress corrosion cracking. They are similar to austenitic stainless steels but are hardenable and of much higher strength. Their cost per pound is seven times that of carbon steel.

9.7 INCOLOY 825

Incoloy 825 is used in extremely corrosive environments such as caustics over 200°F. It has very good resistance to chloride cracking and is resistant to pitting and intergranular corrosion. It is 40% Ni and 20% chromium, with typical tensile strength, 85 ksi hardness, and 160 BHN. Incoloy 925 is comparable to 825 but is high strength and used for downhole hardware and tubular products. Its cost per pound is 13 times that of carbon steel.

9.8 INCONEL 718

Inconel 718 is a nickel-based superalloy that is 55 Ni and 20 Cr. It is used when there are corrosion problems and high strength at higher temperatures is required. At room temperature the tensile strength is 185 ksi tensile and the yield is 150 ksi; at 1400°F it is 166 ksi tensile strength, 137 ksi yield, and is readily welded. It has good low-temperature impact strength properties (20 ft-lb at −320°F). Its cost per pound is 20 times that of carbon steel.

9.9 STRUCTURAL STEEL

Structural steel is a low-carbon steel with 0.27% C maximum. It is a general-purpose mild steel used in general-purpose structural and noncritical applications. I-beams, channel sections, and other rolled shapes, plates, and bars are usually made of ASTM A 36 material. It has 70 ksi tensile strength and 36 ksi yield and is easily welded.

9.10 ALL STEELS ARE NOT THE SAME

When shafts fail, one of the first things that is usually proposed is a higher-strength steel, especially when a failure analysis has not been done. An outside shop that is making the shaft may recommend an alternative steel. By now it is probably obvious that when you get one property, such as an increase in tensile strength, you may be giving up another, such as corrosion resistance. Heat treatments are very important, and simply ordering that the new shaft be made of 4340 may not result in the properties desired. Table 9.2 shows some examples of how the properties change on typical shafting materials, with various heat treatments. Compare the yield strength of a 1-in.-diameter 4340 with that of a 12-in.-diameter forged 4340. It is about half. Make sure that you use the correct specifications for the application.

9.11 USEFUL MATERIAL PROPERTIES

The data in Table 9.3 have been useful when more detailed information on the materials is unavailable.

TABLE 9.2   Effect of Heat Treatment on Steels

Material	Tensile Strength (ksi)	Yield (ksi)	BHN
1020: 6-in. hot-rolled	60	35	120
1040: 6-in. hot-rolled	84	46	160
4140: 1-in. hot-rolled, oil-quenched at 1200°F	145	125	293
4140: 6-in. hot-rolled, oil-quenched at 1200°F	108	80	220
4340: 1-in. hot-rolled, oil-quenched at 1200°F	150	130	302
4340: 6-in. hot-rolled, oil-quenched at 1200°F	125	100	250
4340: 12-in. forged, normalized at 1200°F	95	70	220

TABLE 9.3   Material Properties

Material	E (lb/in^2 $\times 10^6$)	G (lb/in^2 $\times 10^6$)	Tensile Strength (lb/in^2 $\times 10^3$)	Yield (lb/in^2 $\times 10^3$)	Shear (lb/in^2 $\times 10^3$)	Endurance (lb/in^2 $\times 10^3$)
Aluminum						
2014-T6	10.6	4	68	60	39	20
6061-T6	10	3.8	38	35	24	17
Bronze	15	6.5	100	—	70	32
Cast iron						
Malleable	26	8.8	50	32	49	32
Nodular	24	8.8	60	45	—	—
Magnesium						
AZ80A-T5	6.5	2.4	55	38	24	16
Titanium						
5AL, 2.55N	17	6.2	115	110	100	69
Steel, building ASTM						
A7-61T	29	11.5	60	33	17	30
SAE grade 5 bolt	29	11.5	120	80	45	60
SAE grade 7 bolt	29	11.5	133	115	58	66
SAE grade 8 bolt	29	11.5	150	130	65	75
Steel castings	29	11.5	80–175	40–145	20–72	32–70
Stainless 17–7, 1/2 hard	28	12.5	125	78	—	44
Many woods	1.4		1000 lb/in^2	—	1000 lb/in^2	—
Concrete, maximum	3000 lb/in^2	—	350 lb/in^2 tension 3500 lb/in^2 compression	—	1800 lb/in^2	—

9.12   HEAT TREATMENTS

With heat treatment, steels may be made harder or softer, stresses induced or relieved, mechanical properties increased or decreased, along with other changes, such as in structure and machinability. Some of the terms used and their effects are described next.

Normalizing   The steel is subjected to uniform heating at a temperature slightly above the point at which grain structure is affected, known as the *critical point*. This is followed by cooling in still air to room temperature. This produces a uniform structure and hardness throughout.

Annealing   This consists of heating and holding the steel at a suitable temperature, then allowing it to cool slowly. Annealing removes stresses, reduces hardness, and increases ductility.

Stress Relief Annealing   Stress relieving is intended to reduce the residual stresses imparted to steel in operations such as welding or forming, where residual stresses can approach the yield strength of the material. It generally consists of heating the steel to a suitable point below the critical point, followed by slow cooling.

Quenching   Steel is heated to above the critical range, then hardened by immersion in an agitated bath of oil, brine, or caustic. Quenching increases tensile strength, yield point, and hardness. It reduces ductility and impact resistance.

Tempering   The steel is reheated after quenching to a specific temperature below the critical range, then air cooled. It is done in a furnace, in oil or salt baths, at temperatures from 300 to 1200°F. Maximum toughness is achieved at the higher temperatures. Low tempering temperatures produce maximum hardness and wear resistance.

9.13   FAILURE MODES OF SHAFTS, BOLTING, STRUCTURES, AND PRESSURE VESSELS

In this section we show in sketch form (Tables 9.4, 9.5, and 9.6) various failures and what to look for. Whereas Section 2.25 illustrated a generalized failure pattern, in this section we review actual failures experienced by the author, which often included a metallurgical analysis. Photographs were available on many of these failures, but the observations are much clearer in sketch form. Also, some of these failures were sketches made by the author when at the site, observing the failure debris. It takes a trained metallurgist and the necessary tools to interpret many failures, as the failure surfaces are usually too complex for a casual visual evaluation under magnification. These are shown here because similar patterns occurred many times during the author's career and were useful when troubleshooting.

One thing worth noting on many of these failures is the relatively small final rupture zone. This was all the material that was needed to transmit the load, since the shaft was partially cracked through. Many shafts are overdesigned, which as you can see, is beneficial when you aren't aware of all possible causes. In some of the sketches, the failure regions are described as velvety smooth, and others as coarse. The smooth regions are usually associated with the rubbing of the fracture surfaces as the crack propagates due to the cyclic stresses. The coarse region is usually associated with the final fracture.

Also mentioned are *beach marks*, which are usually elliptical patterns about a flaw or defect. These represent a heavy load point during operations such as startups or a shock load on an axle. These lines usually are not obvious under high cycle fatigue. This is because the cycles are close together, such as in a rotating shaft in bending. A metallurgical laboratory with scanning electron microscopy may be able to make sense of them.

Brittle fractures or brittle material failures can be a little more obvious. Often, there are ridges radiating from the failure origin or flaw. Also, in brittle fractures there can be chevron patterns that point toward the failure region.

TABLE 9.4   Shaft Failures

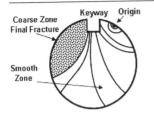

Failure:
Rotating, bending of a motor shaft

Cause:
Stress concentration on shaft surface due to a grinding gouge. Bending due to belt drive provided alternating stress.

Material:
4140 steel

Observation:
Very distinct origin. Goes from smooth to rough moving away from the origin.

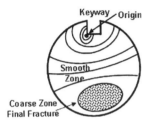

Failure:
Rotating, bending of an output shaft

Cause:
Loose key and very sharp keyway radius. Hard startups propagated crack.

Material:
Carbon steel

Observation:
Obvious origin. Shaft stress was low since it took a long time to fail, evidenced by large fatigue and small fracture zone.

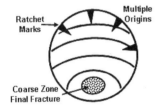

Failure:
Rotating, bending of gearbox shaft

Cause:
Spline on output end locked up and caused high bending load at snap ring groove.

Material:
4140 steel

Observation:
Most obvious was off-center final rupture, which indicated load in one direction.

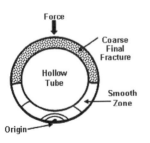

Failure:
One-way bending of a tube with no rotating bending

Cause:
Heavy operating loads in one direction

Material:
Carbon steel, Brinell hardness 215

Observation:
Surface imperfection at origin combined with one-way bending

TABLE 9.5   Bolt Failures

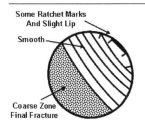

Failure:
Tensile failure of bolt

Cause:
Bolt became loose and saw full tensile cycle load.

Material:
Grade 8 material

Observation:
Smooth appearance with some ratchet marks at the origin of failure. Almost straight lines progressing from the origin to the coarse final failure zone.

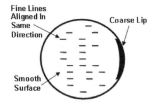

Failure:
Shear failure of bolt

Cause:
Bolt was clamping coupling halves that experienced high shear loads. Torque overload caused direct shear in body of bolt.

Material:
B7 bolt material

Observation:
Smooth appearance of fracture surface with only a small lip at the outer edge. Under 6× magnification, grain seems to flow in direction shear occurred.

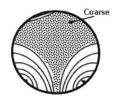

Failure:
Bending fatigue failure of bolt

Cause:
Bolt was used to secure blades to agitator. Bolts came loose in service, which caused bending loads as blades were struck by product.

Material:
17-4 PH bolt material

Observation:
Looked similar to shaft bending failures. Had multiple failure sites with smooth beach marks.

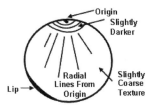

Failure:
High-strength coupling bolt failure on rotating equipment

Cause:
Stress corrosion failure of bolt due to environment. Wrong material selected.

Material:
High-alloy steel, 260 ksi ultimate strength.

Observation:
No cyclic loads on bolt, as bolts tightened sufficiently. Several bolts in assembly failed in first thread. Surface was slightly coarse, not velvet smooth, with ridges radiating from origin. Darker semicircles appeared at the origin.

TABLE 9.6   Structure and Pressure Vessel Failures

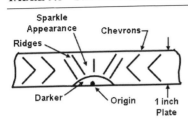

Failure:
Burst test on 1-in.-thick wall pressure vessel at low temperature

Cause:
Brittle fracture was forced to occur in an older vessel with built-in attachments and flaws.

Material:
Carbon steel with 85 ksi ultimate strength

Observation:
Failure occurred at an attachment weld. Appearance is of ridges radiating from the origin and V chevrons pointing in the direction of the origin or flaw. The surface had a "sparkling" appearance, like quartz crystals in quartz.

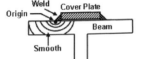

Failure:
Fracture of a vibrating screen structure originating at a weld

Cause:
Misalignment caused high bending distortion of the structure and cyclic stresses in the welds.

Material:
Structural I-beam with welded-on attachment cover plate

Observation:
Fatigue failure started at the toe of a good weld and progressed through the structure. The sketch represents the appearance of the failure. Beach marks were from startup loads, and the velvety smooth appearance was from the 1000-cycle/min vibratory nominal stress of about ±6 ksi.

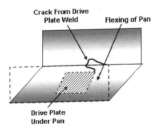

Failure:
Crack originating at weld on vibrating drier drive plate

Cause:
Misalignment caused bending flexing of C-section $\frac{3}{32}$-in.-thick formed pan transmitted through the drive plate.

Material:
316 stainless steel

Observation:
A wandering type of crack that grows very slowly due to the low propagating stresses involved. Cyclic stresses were less than ±5 ksi in this case.

Cracks in sheet metal or thin plates can sometimes be arrested temporarily by drilling a small hole at the ends of the propagating crack. This is done on aircraft structures to slow the growth until a permanent repair can be made. The theory is that the $\frac{1}{16}$-in.-diameter hole drilled at the tip of a crack produces a known stress intensity of 3 or 4, whereas the tip radius of a crack tip might be many times higher than this. When the local stresses are low, the small hole is enough to lower the stresses to a point where the crack won't grow anymore. The real problem is making sure that you found the tip of the crack when you drill the hole. When a crack is not straight from one point to another but wanders around, there is usually only a small driving stress.

9.14 FRETTING CORROSION

Wear was discussed in general in Section 2.34, Fretting is a form of wear and was discussed briefly in Section 2.25.7. Fretting has been experienced by the author on splines, bearing outer and inner race housing fits, gear shaft fits, shaft coupling fits, geared coupling teeth, bolted joints, aluminum riveted joints of aircraft structures, keys and keyways in shafts and hubs, and universal joint pivots. Fretting has not been limited to steels and has also been observed on high-speed breaker pivot points on a polymer material.

Fretting is caused by small relative motion, called *slip*, between two surfaces that were probably not intended to move. The slip amplitude is typically less than 0.002 in. (50 μm) and can be much smaller. There doesn't appear to be a slip amplitude that will stop fretting once it has occurred. There can be a threshold amplitude below which fretting will not initiate [41,42]. This rubbing in steels results in fine wear particles that usually oxidize and have a powdery red appearance. When mixed with oil, it has a blood red appearance. Usually, a hard tenacious brown scale is also present. Aluminum fine wear particles have a charcoal powdery black appearance.

Although lubrication appears to help reduce fretting of splines and geared couplings, it can also accelerate fretting in other cases, such as hubs and gear fits. This is because the lubricant allows more micromotion and thus more fretting. Fretting does cause destruction of the surface from the wear, but in the author's experience it is the effect on the fatigue life of the shaft or part that has been of the most concern.

The fatigue strength or endurance limit under fatigue test fretting conditions can be reduced by 50 to 70%. This information is academic in operating machinery since it all depends on when the fretting started and how fast it progressed. Some shafts have failed from fretting after 10 years of service. It took that long to degrade the surface to the point where fatigue cracks could grow through the shaft. Others have failed a few months after having been put into service.

Fretting causes microcracks which are believed to reduce fatigue life. This is not hard to imagine since 90% of fatigue life goes into just starting a crack. Since fretting produces these cracks, the component's life is reduced. This is shown in Figure 9.4. Notice how the fatigue life is reduced as the initial crack depth increases. With the smallest of cracks and a sizable cyclic stress, the life starts to decrease immediately.

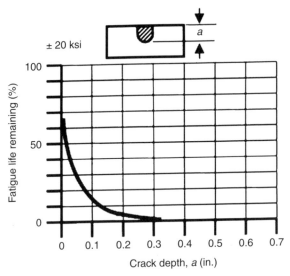

FIGURE 9.4   Initial crack size and life.

Preventing fretting is difficult. Designing to minimize fatigue can be done, and some methods are suggested in the references. Tightening fits on couplings and gears and tolerances on bearing races may work but could also cause other design concerns, such as overstressing or preventing growth, as in the case of the outer race. The author has found fretting to be one of the most difficult problems to solve. Tightening tolerances on fits has been helpful but may not always work. Tightening key fits in keyways and hubs has had mixed results. In some cases it works and in others it doesn't; it depends on what is causing the micromotion. When it is the flexing of a shaft during operation, it might depend on the magnitude of the cyclic loading, so just tightening the fit might not be successful.

In Section 2.34, the following wear model was discussed in the form of the wear rate:

$$\frac{\delta}{t} = \frac{370K\sigma V}{\mathrm{BHN}} \quad \mathrm{in./yr}$$

Recall that δ is the wear depth, t the time for the wear depth to be reached, K the surface condition (i.e., lubrication and friction effects), σ the stress with which the parts were pressed together, V the velocity of the rubbing, and BHN the hardness of the surface. Since the amplitude of the sliding $S = Vt$,

$$\delta \text{ is proportional to } \frac{K\sigma S}{\mathrm{BHN}}$$

So wear is directly proportional to the surface condition K, contact load σ, and amplitude S, and is inversely proportional to the hardness BHN. This says that wear

δ, once started, proceeds linearly with increasing contact load σ and increasing amplitude S, and reduces as hardness is increased.

We mention that increasing a fit (i.e., increasing σ) can reduce fretting, and the equations says that it will increase wear. The reason is that tightening the fit also reduces the amplitude S; that is, the fretting motion is also reduced. Fretting obeys this wear equation; however, the problem is what K value and amplitude S to use. For fretting where the fixed and sliding materials are similar, and with poor lubrication or good lubrication, K can vary from 10×10^{-6} to 100×10^{-6}, respectively [23, p. 503]. For design modifications or troubleshooting purposes, the equation will show directionally how fretting can be reduced. The question always is if that will be enough to solve the fretting problem. Obviously, if no fretting was occurring until the machine loads were uprated, and then fretting started, valuable information is available. An analysis similar to that in Section 2.35 might be applicable.

Designs are available to reduce or eliminate fretting. Nitriding, which increases the surface hardness of splines, and frequent lubrication by greasing have been used to reduce the fretting of gearbox splines driving extruder screws. Here the fretting micromotion is due to the free-floating screws, which bend and misalign very slightly. This micromotion at the splines resulted in the fretting. Fretting was never eliminated completely, only reduced to a tolerable level.

Tightening keyway fits, as mentioned in Section 2.25.7, aided in reducing shaft fatigue failures but was not always successful. This could have been because the amplitude of the micromotion S was not reduced, or something unknown occurred to loosen the key, such as operating impacts.

Cyclic loading at stress concentrations can result in fretting due to the strain and micromotion experienced. Consider the gear–shaft with a fluctuating rotating bending load shown in Figure 9.5. The gap on the end of the shaft, represented as δ_H, is the stretch of the bottom fiber due to the shaft deflection δ_V. To get some idea of the magnitude of this, consider that δ_V is due to a load in the center of a simply supported shaft: From geometry, the slope in radians at the end of the shaft is

$$\theta = \frac{2\delta_H}{D} \qquad \text{where } \tan\theta \sim \theta$$

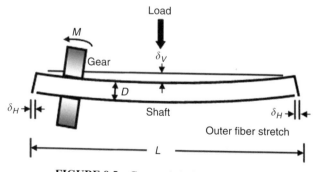

FIGURE 9.5 Gear and shaft with fretting.

From the beam deflection end slope in terms of the center deflection,

$$\theta = \frac{3\delta_V}{L}$$

Making the substitution and solving yields

$$\delta_H = \frac{3}{2}\frac{D}{L}\delta_V$$

A constant bending load such as a gear loading will result in this axial displacement δ_H. Every time the shaft rotates about its centerline, the outer fiber stretches and contracts, with the stress going from tension to compression. If the gear is held to its slope due to the applied loads, fretting at the bore could occur. The axial motion would be δ_H and could then result in fretting.

Consider a shaft of diameter $D = 4$ in. and length $L = 24$ in., deflected in the center by $\delta_V = 0.001$ in., which was caused by a center load. How much would the shaft move in and out if it was free to slide at the end supports as it rotated?

$$\delta_H = \frac{3}{2}\frac{D}{L}\delta_V$$

$$= \frac{3}{2}\left(\frac{4}{24}\right)(0.001)$$

$$= 0.00025 \text{ in.}$$

This is within the fretting range that could be expected; however, it is also within the deflection range that many shafts will experience. Not many shafts experience fretting, so there must be many other factor that prevent fretting from occurring. Notice that if another gear was at the end of the shaft, it would see some of this axial motion and might fret at the shaft–bore fit. The relative motion between the shaft and the bore fit would need to be determined, which is not an easy problem to solve. At the end of this section a semiempirical approach is provided.

Methods for minimizing fretting at stress concentrations are shown in Figure 9.6. Basically, the stress concentration is minimized where the rubbing would occur.

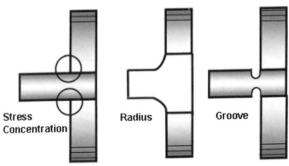

FIGURE 9.6   Minimizing fretting at fits. (From [20], p. 324.)

FIGURE 9.7   Inner race fretting at shaft.

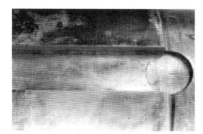

FIGURE 9.8   Fretting on a gear shaft and keyway.

Figure 9.7 shows the inner race, bore-to-shaft fit. The close-up shows fretting wear caused by a bad bearing and a broken gear tooth. The shaft did not fail, but the fretting indicates the loss of material. The fretting was evident only on the side of the gear shaft with the failed bearing. Micromotion from the failed bearing vibration caused the fretting. The unit had been in service for seven years.

Figure 9.8 shows fretting on the outside diameter of a bull gear shaft but not in the keyway. The gear had a loose fit to allow axial sliding and alignment of the gear set. Also shown is a badly fretted keyway, which was caused by a loose-fitting key.

When shafts have failed due to fretting, especially in keyways, many companies remove the gear or extruder flights periodically and dye-check the shaft with liquid penetrant in the key region. Cracks should be welded using an approved technique, or the shaft replaced if badly cracked when lost production is a major concern.

Keys should be a snug fit in the keyway (i.e., a snug push-in by hand). Keys should be the full length of the keyway and have rounded ends to fit the keyway, and the keyway should have a radius at the bottom. Keys should have an angle or radius on the bottom so they don't interfere or cut into the shaft keyway radius.

The following semiempirical screening procedure has been used to determine if fretting was probable on the shaft under the shrunk-on thrust collars of large, high-powered gearboxes. The method has been used to examine only six cases in actual service at the time of this publication, but since this type of information is difficult to obtain, it is worthy of mention. Gill [43] has come up with a model for shrunk-on fits which states that once the local alternating stress at the fit location is greater than the joint friction capability, the likelihood of relative motion, and thus fretting, is high. In equation form this is stated as

$$\sigma_{alternating} > \mu p_{contact} \qquad \text{fretting possible}$$

Here μ is the coefficient of friction between the fit and is usually taken as 0.12 and $p_{contact}$ (lb/in^2) is the net contact pressure due to the press fit (*net* means after the centrifugal force loss-of-fit effect has also been considered). The alternating bending stress on the surface of the shaft is $\sigma_{alternating}$ (lb/in^2), whose value is

$$\sigma_{alternating} = \frac{Mc}{I}$$

This shaft bending equation is explained in Section 2.2.3.

Gill reports that, using this screening tool, a shaft with a full load shaft bending stress at the shrink-fit junction in excess of 3500 lb/in^2 fretted with a contact pressure of 20,000 lb/in^2:

$$\sigma_{alternating} > \mu p_{contact}$$
$$3500 > 0.12(20,000)$$
$$> 2400 \qquad \text{fretting possible}$$

In only one case, where the alternating stress value was 1700 lb/in^2 and the unit was not operating at full load, was fretting not evident.

$$\sigma_{alternating} > \mu p_{contact}$$
$$1750 > 0.12(20,000)$$
$$> 2400 \qquad \text{fretting not evident}$$

Although the number of verification data are meager, Gill is continuing to collect data, and the model certainly has merit. It seems entirely possible that when the local axial frictional forces trying to hold the shaft and collar together are overcome by alternating bending forces, relative motion and thus fretting could occur. This simple relationship also provides an explanation of why increasing the fit pressure is not always successful. The bending stresses may be just too large. It also explains why lubrication can sometimes increase fretting rather than reduce it, since decreasing μ will reduce the right-hand side of the equation.

10

MECHANICAL SYSTEM MODELING, WITH CASE HISTORIES

Throughout this book, equations have been developed and used to help troubleshoot actual problems. Examples and case histories were employed to illustrate use of the equations. In this chapter, use of this analytical problem-solving technique is expanded. The actual case histories resulted in substantial insight into the causes of failures.

10.1 SIZING UP THE PROBLEM

Very rarely in troubleshooting are failures due to blatant design errors. Techniques for designing equipment have usually been good and are getting better. For critical equipment, design codes and industry standards usually produce adequate designs. For noncritical applications, the manufacturers internal design philosophies have allowed them to remain in business. Scale-up on equipment is usually accomplished based on historical experience and analytical methods.

Bearings, gears, bolts, shafts, and other components don't usually fail because the equipment builder didn't calculate the loads and stresses correctly. Pressure vessels and piping don't leak because the designer didn't calculate the thickness or stresses correctly. They fail to perform their intended function because something has changed from what was originally designed. The question is: What has changed? More specifically, how can an analytical model help?

Consider the following example. A motor drives a gearbox that drives an extruder screw. This extruder screw fails in torsion. The extruder has been in operation for 15 years and this is the first such failure. The model considers the system geometry, motor horsepower, and shaft speed, so the stress in the shaft is known. The calculations indicate that under the original design conditions, the shaft should not have failed. Indeed, it has operated for 15 years without failing.

Since the geometry and speed have been constant, one would have to question the horsepower. A motor is often upgraded to higher horsepower duty without re-rating of the entire system. In a case such as this, either the analytical model is wrong or something has changed from the original design. In this case history, the motor amperage, should be checked.

When a good representation of the system is available in equation form and the equation results do not agree with the failure mode, valuable information can be produced as to the cause. For example, if a shaft is failing in torsion but the model calculates that the stresses are acceptable, the possibility of torsional impact or torsional vibration should also be reviewed. With the cause known and verified, the true power of the model can be utilized. For this example the shaft stresses or material properties needed for a long life at higher horsepower can be discussed with the manufacturer.

One precaution on design is when machinery or equipment is purchased that is very cost competitive. Reducing costs can result in less metal in critical areas. All too often on new installations used on low-budget projects, the lowest-cost equipment is purchased. Although the project cost may be low, the life cost of the equipment to the owner can be high. This is usually due to lost production opportunities and maintenance costs. Paying more for good-quality equipment is often the correct decision for long-term, low-maintenance, high-service-factor performance.

Determining What to Include in the Model   There is no use building a model of a system unless it produces results with which you can work. When the model you have developed has only the equipment geometry included, and the geometry has been set by the manufacturer or process, your model is not going to help. What are you going to modify? However, if you have included some type of loading and that loading can be changed, you have included a potential solution. Loading causes stresses, displacements, and wear, and excessive stresses, displacements, and wear result in failures.

Theoretically, you should include as many of the variables as are available. Practically, you should include as many as you are capable of handling. The mathematics can get out of control quickly, and sometimes we develop equations we have no hope of solving. It is sometimes better to start with simple models. There are a couple of good reasons for this. First, you can always try a simple model and add to it if more variables are needed. If the results prove reasonable and verify the cause, you are done and there is no need for a complex model. Second, simple models are easy to understand and explain to others. Troubleshooting usually involves explaining the cause and potential remedy to others. It certainly is easier to do when you fully understand the model you have developed. This is especially true when you are explaining the results to those who do not have a technical background.

The equations you will need in your toolbox will vary with the engineering discipline you are addressing. In mechanical engineering the basics of mechanics, such as statics and dynamics, will be invaluable, as will be the basic stress equations. Energy relationships will also be useful. Excessive loads due to pressures, bolting, belt tension, restrained thermal differences, hydraulics, misalignment, imbalance, vibrations, impacts, and increases in horsepower all are major contributors to failures. The case histories examined in this book will have one or more of these loads contributory to the failure. If excessive, these loads, can have a dramatic effect on a component's design life. For example, in roller bearings, doubling the applied load can reduce the design life by 90%. Impact loads such as those due to surge and water hammer can also cause rapid failures.

As mentioned, it is not the original design loads that usually cause failures, but unexpected deviations from these design loads. When building a model, this fact can be used in our favor. Consider yourself to be the equipment designer. What would you do to cause the equipment to fail as it did? In the case of a bearing failure, what would have to happen for it to fail in overload? The radial load could have increased. This could be due to excessive gear load reactions, V-belts being too tight, or misalignment of shafts. An increase in shaft temperature could cause excessive preload in the bearing, and that will result in a rapid failure. A good equipment design would have some design margin included to account for these contingencies. Unfortunately, re-rates, operating upsets, and operational or maintenance errors can use up the original design margin. When it does, this can result in the change that caused the failure.

Vibration of systems requires an energy input source, and to reduce the vibration amplitude you must determine what is causing the energy and how to reduce it. An understanding of the resonance of linear or torsional systems is required to do this. The model you construct should contain energy sources and/or geometry, such as shaft diameters, lengths, and masses. These parameters can be modified to reduce the vibratory amplitude and thus reduce or change the frequency and the loading.

Failure due to excessive wear will require parameters that affect wear. It is also important to understand the wear mechanism that is causing the wear. Abrasion, adhesion, and corrosion are some of the more common types. For abrasion and adhesion, understanding the various wear relationships is necessary. These usually contain the applied load, velocity or speed (rpm), hardness of the surface, lubrication conditions, and type of material. The model should contain these parameters, and since wear is usually defined as a loss of material, a volume loss should be included. Here one of the parameters that can be changed to reduce wear is the surface hardness. This is done by selecting a more effective hard-surfacing method to minimize the volume lost.

Impact failures, such as the sudden closing of valves, causing hydraulic impact or water hammer, or an impeller being struck by chunks of product, usually include the time for the closure to occur. In many cases, such as mixer impact, equations are available that include impact factors. These factors have been determined from full-scale testing on instrumented agitators. For any meaningful analysis such equations should be used in the modeling. In this way the impact factor can be solved for and could be used to identify the cause of failure.

It is quite possible that after reviewing the data, it may be decided that the problem is too complex to try to simplify. Sometimes this happens after our best attempt at building a model fails. It may be appropriate at that point to consider more detailed methods of solution, such as the finite-element method or other approaches. It may even be cost-effective to use a consultant who specializes in these areas. One thing is certain: If you have attempted to model the system, you will have a much better understanding of the problem and be able to discuss other possibilities with a consultant in an intelligent manner.

Analytical Modeling Procedure   When you have no idea what the failure mode is, you are not going to be able to model the problem effectively. It is important to know if you have a low- or high-cycle fatigue problem, a wear-related problem, a sudden failure or corrosion-related problem, or a failure because the wrong material or heat treatment was used. A good materials laboratory report can supply this type of information if you supply the failed component. Remember not to clean the part, wire brush the surfaces, or damage the fractured surface. Even rust and corrosion debris on the surface can contain valuable information in determining the failure mode.

When the fracture surfaces have rubbed together, the evidence can be destroyed and the materials laboratory may not be able to determine the cause of failure. This can happen when a shaft shears through and continues to rotate. The shaft ends can rub together, destroying the surfaces. In this case all that can be done is to assume a reasonable failure mode with the data available, and model it.

When the failure mode is clear enough, both the cause and the remedy are often obvious. For instance, if a fatigue failure occurs on an unsupported pipe in vibratory service, the solution is to add a gusset to the connection. This eliminates the weak link and solves the problem. Probably 95% of failures are handled in this manner. It's the 5% that aren't this easy that require modeling.

The results should agree with the failure mode. If the failure mode was low-cycle bending fatigue and your model is an impact model, the wrong mode has been evaluated and you will have to build a model that considers cyclic bending loading. This is why it is always prudent to have the failure mode known before the model is built. It is both embarrassing and a waste of time to find out that you have analyzed for a failure mode that hasn't occurred. This is one of the dangers in developing a model without knowing the failure mode. One can get so involved with the mathematics and the many variables involved that any practical simplified solution is lost. Having a laboratory report will keep the analyst honest and bring one back to reality.

With a solution that agrees with the failure mode, the results can still say that the component should not have failed. But it did! So either the laboratory report is in error or the model is showing that the loads are not high enough to cause the failure. The question, then, is what could have caused the loads to be higher. When we determine this, we may have determined what caused the equipment to fail.

When no single reason for the failure is obvious, it is sometimes prudent to address several potential causes, just to be on the safe side. The consequence of a failure will drive how far you take this. If a small belt-driven shaft for a testing machine keeps breaking in bending fatigue, a higher-fatigue-resistant material, elimination of stress raisers, and better belt-tensioning procedures may be necessary. This would be especially true if the testing was done as a quality check and interrupted normal production of a product line. If there were no lost opportunity credits and the machine was not critical, perhaps a better belt-tightening procedure was all that was needed.

Finally, there are failures for which we really have no clue as to why they occurred. For example, consider the case where a system went down at 3:00 a.m., was repaired, and was put back in service. The failed parts were thrown away and the computer went down at that instant, so no failure data were taken. It happens. You don't know if these types of failures will happen again or how bad the failure might be next time. This usually makes everyone quite nervous. About all you can do with these types of failures is to monitor, monitor, and monitor. Determine the type of data you will need to identify the cause if another failure occurs, and provide the instruments to capture these data.

At one time a reactor with a propeller-type bottom-entry agitator would self-destruct periodically. It seemed rather random and would stop production when it came down. Modeling, analyzing the worst flow conditions, did not produce a clue. The failure mode of the shaft and blades was sudden impact. A continuously monitoring ampere reading and vibration monitoring device were installed along with other instrumentation. A week later amperage and vibration spikes occurred during a specific part of the operating cycle, but nothing failed. It happened again five days later at the same point in the operating cycle, and the flow went to zero. The shaft had failed.

A review of the reactor's operating history indicated that at that point in the cycle the reactor is being washed. Chunks of product had built up on the reactor walls due to operation problems and fell onto the spinning blades. By the time the reactor was opened for inspection, the chunks had dissolved and the evidence was gone—sort of like an ice bullet being used in a murder. The damage is there but the evidence has disappeared. Monitoring solved the mystery. The model had predicted that such a high amperage could result in a broken shaft. The problem was that no one could come up with a reason why such high power (amperes, as volts were constant) use would produce a failure.

The Funny Look Test When your model is complete and verification cases have been performed, things can still not be right. As mentioned earlier, sometimes we are so engrossed in an analysis and the mathematics involved that we can lose sight of what we are trying to do. We may be doing analysis for the sake of analysis. We can sometimes force an incorrect cause.

A method that has been found useful to check an analysis is to put it aside for several days, work on something else, and then come back to review the results. Things just might not seem right to you: that is, it doesn't pass the "funny look test." Some indications of this are:

- The results don't seem to make sense; for example, the model predicts a fatigue bending failure and the laboratory results say that it is an impact failure.
- It doesn't seem to agree with what you have seen in the past, and your instinct rejects it.

- Simple calculation checks of the model are off by a factor of 10 or more.
- No one believes your results.
- Your results violate one of the fundamental laws of nature. For example, your model predicts that the wind will cause a tower to deflect against the wind direction.

If any of the above occur, it would be wise to perform a fundamental review of your model and to discuss it in detail with a colleague who is familiar with the failure you are trying to solve.

10.2 CASE HISTORIES

Although some of these case histories were followed up with more elaborate analysis methods, most were usually sufficient to provide valuable results to help solve a problem. Most were of the high-impact type, so they resulted in production limitations and received considerable high-level management attention. All of these case histories required the input of many talented people to obtain the results. Gathering the failure data and implementation of the results are usually the most difficult parts of problem solving. In seminars and articles, the author has heard how easy these case history problems look to solve. Most were not, and although an analytical case history can be written up on a few pages, the problem may have taken weeks, with many team meetings, to solve and implement. Murder mysteries are solved on a television show in one hour, but in real life it might take months or years. The same is true for case history troubleshooting. The drudgery of obtaining the facts is left out. Case histories are presented here because they contain much of what has been explained throughout the book. Although the reader may not experience exactly the same problem, the analytical models shown can usually be adapted in some way to address the problem being worked. The case histories are presented under the following categories: (1) failures caused by excessive loads, (2) failures caused by wear, and (3) failures caused by thermal loads.

10.3 FAILURES CAUSED BY EXCESSIVE LOADS

10.3.1 Case History: Agitator Bolt Failure

Agitators are pieces of equipment used to mix and blend fluids. They can be mounted on top, under, or on the side of a vessel and include a rotating shaft connected to several blades. The blades can be from a few inches to several feet in diameter and are usually bolted to a hub. This case history examines blade bolt failures on a top-mounted agitator, shown in Figure 10.1. When the 20-ft-diameter slurry tank was entered for cleaning, it was not unusual to discover a blade on the bottom of the tank. These blades have the potential of puncturing the tank bottom if they happen to hit corner end down. Most of the time the unbalanced forces due to the

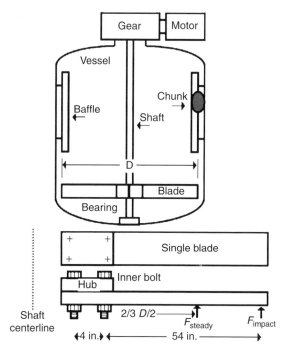

FIGURE 10.1   Top-mounted agitator impacting.

missing blade also wear out the steady bearing and a seal leak develops. Sometimes the broken bolts that held the blade were found, and sometimes they were not and later ended up wrecking a pump somewhere else in the circuit. Pump suction strainers were not used in this service because of the composition of the slurry and the potential for plugging. Any of these scenarios would result in extended downtime and lost production. A metallurgical examination of the badly battered bolts indicated that two different modes of failure were present. A sudden bending failure was one mode, and a fatigue failure starting in the bolt threads was the other.

One of the most important things that an engineer can do when troubleshooting any equipment problem is to talk to others who are very familiar with the equipment. This is not always the manufacturer, especially when troubleshooting. Some of the best sources are mechanics, operators, and technicians who have been associated with the equipment. Discussing this particular failure with them generated the following information:

- One of the four nuts per blade was sometimes found loose.
- Basketball chunks of product were found wedged between baffles.
- Even when "double nuts" were used, some of the bolts were found loose.
- Bolts were found bent like pretzels.
- The bolts were tightened with hand tools.

There is a lot of good information here. With the materials laboratory information on the modes of failure, a theory can be formulated on what is happening. It was hypothesized that the bolts became loose because a blade periodically struck a chunk of material that became dislodged from a baffle. This is sort of like hitting a baseball with a bat. Your hand feels the impact, just as the bolts would.

There are quite a few variables in this case history, and to get a handle on them, an analytical model was built. The model was to be used to determine what the external load and preload was on the bolts and what could make the bolts become loose. The model was then used to identify improvements.

The tip velocity of the agitator blades is

$$V = \frac{\pi D(\text{rpm})}{60} = 282 \text{ in./sec}$$

The impact force on the agitator blade is caused by a chunk of product of weight $W = 10$ lb, with a spring constant $K = 100$ lb/in. being hit by the blade at the tip. K was determined by placing a 200-lb weight on the product and noting that it deflected 2 in.

$$F_{\text{impact}} = V\left(\frac{KW}{386}\right)^{1/2} = 282\left[\frac{100(10)}{386}\right]^{1/2} = 454 \text{ lb}$$

In addition to the impact load, there is the steady-state load due to the blade moving through the product. The total force per blade is derived from the horsepower equation:

$$\text{hp} = \frac{T(\text{rpm})}{63,000}$$

$$F_{\text{steady}} = \frac{63,000(\text{hp})}{(\tfrac{2}{3})(D/2)(\text{rpm})}$$

$$= \frac{63,000(75)}{(\tfrac{2}{3})(60)(45)}$$

$$= 2625 \text{ lb}$$

$$F_{\text{blade}} = \frac{F_{\text{steady}}}{\text{number of blades}} = \frac{2625}{4}$$

$$= 656 \text{ lb}$$

Moment summation about the innermost two bolts:

$$4(2)P_{\text{bolt}} = (454 + 656)(54)$$

$$P_{\text{bolt}} = 7500 \text{ lb}$$

Preload in bolt with $T = 150$ ft-lb torque; bolt diameter $d = 1$ in.

$$F_{preload} = \frac{60T}{d} = \frac{60(150)}{1} = 9000 \text{ lb}$$

Stretch in the bolt due to preload with length $L = 4.5$ in. and bolt area $A_{bolt} = 0.785d^2$:

$$\delta = \frac{F_{preload}L}{A_{bolt}E}$$

$$= \frac{9000(4.5)}{(0.785)(30 \times 10^6)}$$

$$= 0.002 \text{ in.}$$

Stress in bolt due to $F_{preload}$:

$$\sigma = \frac{120T}{d^3}$$

$$= \frac{120(150)}{(1)^3}$$

$$= 18,000 \text{ lb/in}^2$$

What the model shows is that the impact force plus the steady-state force effect at the bolts never really exceeded the bolt preload. When the bolts are kept tight, a fatigue or impact failure should not occur. This is because in this bolted joint the bolt will not see any additional load or stress unless the bolt preload is exceeded. Also, the analysis clearly shows that there isn't much stretch in the bolt, so that relaxation of the threads or embedment of the surfaces could easily cause the bolts to loosen. Relaxation–embedment is expected to be about 0.001 in. for this joint, which is much too close to the 0.002-in. stretch achieved. Once the bolts loosen and the preload is lost, the bolts will see the full effect of any impact. With loose bolts a crack would be likely to occur and the bolts would fail in fatigue. Also, it is evident from the moment reaction point that the inner two bolts are not helping much. That is why they find bolts "bent like pretzels." When the outer bolts fail, all loads is transferred to the inner bolts and the blade bends these bolts. The bolts are a ductile stainless steel with low yield strength, so they bend until they fail in tension.

To address the cause of loose bolts the bolt material was changed from 316 stainless steel with a yield strength of 30,000 lb/in² to Inconel 718 with rolled threads and a yield strength of 150,000 lb/in². This allowed the preload and thus the stretch in the bolt to be increased by a factor of 4. With 0.008-in. stretch and a relaxation of only 0.001 in., the bolts remained tight, even with occasional product impact. Double nuts were continued since it made everyone feel more secure; this was not required, however.

10.3.2 Case History: Loosening of Counterweight Bolts

If there is one rule of thumb that should be applied to minimize bolting failures, it is: "Keep the bolts tight." Loose bolts, or bolts where the operating applied load is

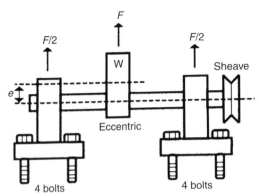

FIGURE 10.2 Forces due to eccentric weight.

greater than the assembly preload, can see the full effect of any cyclic load. Excessive cyclic loading can lead to fatigue failures of bolts. In Section 2.33 we examined static loads on gaskets and how the bolts can loosen because of process upsets and the effect on gasket leakage.

In this case history, the bolts that failed in fatigue were used to secure the pillow block bearings, which were supporting an eccentric weight shaft that drove a vibrating screen conveyor (Figure 10.2). The question was: Were the bolts tightened enough? This was important to know, because when the bolts failed, extensive damage was done to the machine structure. Replacing the bolts without a root cause was not an option because the consequence of a repeat failure was substantial.

The following analysis was performed to see if the operating load was greater than the preload. In Figure 10.2 are shown the forces due to the eccentric weight $W = 100$ lb, $e = 1$ in., rpm $= 750$:

$$F = 28.4We\left(\frac{\text{rpm}}{1000}\right)^2 = 1600 \text{ lb}$$

$$F_{\text{bolt}} = \frac{F/2}{4} = 200 \text{ lb per bolt cyclic load}$$

These were $\frac{1}{2}$-in. bolts torqued to 30 ft-lb, which produces an assembly preload of 3800 lb. With this amount of preload and only a 200-lb cyclic load, the bolt would see little, if any, of the cyclic stress, and a failure in fatigue should not have occurred. A torque wrench had been used, so the bolts were tightened correctly, and there were no other loads. What, then, caused the fatigue failure?

When a bolt is tightened, it stretches like a spring and if the joint relaxes, some of this stretch is lost. When there is no more stretch, there is no more preload and the joint becomes loose. All of the cyclic load then goes into the bolt, and fatigue failures occur quickly. The stretch in these 3-in.-long bolts is

$$\delta = \frac{PL}{AE} = 0.0019 \text{ in.}$$

where $P = 3800$ lb, $L = 3$ in., $d = \frac{1}{2}$ in., $A = \pi d^2/4 = 0.196$ in^2, and $E = 30 \times 10^6$ lb/in^2. This is not much stretch, and typically if the stretch is less than 0.002 in., relaxation of the threads and joint will be enough to cause a loose bolt. The solution for this case was to use a spacer sleeve so that a smaller-diameter, longer, higher-strength bolt could be used. This allowed the bolt stretch to be increased to 0.006 in., which was adequate to keep the bolts tight, even with the vibratory loading.

10.3.3   Case History: Evaluating Internal Thread Strip-Out

When internal threads are damaged in tapped holes, the following question can be expected: "Can we continue to operate with the threads damaged?" Decisions to repair immediately by welding up the holes and re-tapping, or by drilling larger holes and re-tapping for larger bolts, or by using threaded inserts with the same-size bolt can be made by anyone. The tough call is to determine if the unit can be kept operating safely until the next planned downtime without incurring production debits.

Consider a stud in a threaded hole, where, as with a nut, no dilation is possible (Figure 10.3). The stud is steel and the part is ductile steel, which has a yield strength of $\sigma_{yieldpart}$, which is less than that of the stud. L is the engaged length at which strip-out of the threads in the part could be expected, D is the bolt nominal diameter, and P is the load in the bolt.

At complete yield conditions of the thread, the load is pretty evenly distributed along the threads [44], and the shear strength is about 0.577 of the yield strength. The failure stress (σ_{fail}) can be approximated as

$$\sigma_{fail} = 0.577 \sigma_{yieldpart} = \frac{P}{A_{shear}} = \frac{P}{\pi D L}$$

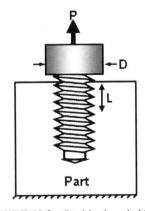

FIGURE 10.3   Stud in threaded hole.

The thread length that will strip out is

$$L = \frac{1.73P}{\pi D \sigma_{\text{yieldpart}}}$$

Knowing the nominal stress (σ_{bolt}) in the bolt due to the assembly torque or stretch and any other loading, the bolt load is

$$P = \frac{\pi D^2}{4} \sigma_{\text{bolt}}$$

This simplifies to

$$\frac{L}{D} = 0.43 \frac{\sigma_{\text{bolt}}}{\sigma_{\text{yieldpart}}}$$

Figure 10.4 graphs this equation along with test data, shown as crosses, from various sources. In this case history a 3-in.-diameter UC–8 TPI (threads per inch) steel stud was threaded into a hot ductile steel vessel flange. The stud bolts were removed to facilitate inspecting a flange. During this procedure, eight of the 31 threads were damaged. The question was: Can the vessel be put back into service and operated or is a long and costly downtime required for repairs?

The studs would be loaded with a hydraulic tensioner to 45,000 lb/in².

$$\frac{L}{D} = 0.43 \frac{\sigma_{\text{bolt}}}{\sigma_{\text{yieldpart}}}$$

$$= 0.43 \left(\frac{45,000}{30,000} \right) = 0.645$$

$$L = 0.645(3) = 1.94 \text{ in.}$$

The threads could be expected to strip out in service if the engagement was 1.94 in. The remaining engagement is 23 good threads or 2.88 in., so stripping out isn't expected. It is interesting to note that the original thread engagement of 31 threads/ 8TPI = 3.875 in. provided an L/D ratio of about 1.3, which is close to conventional

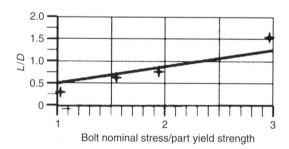

FIGURE 10.4   Stripping strength of threads.

design wisdom, which suggests an engagement length of about $1.5D$ for bolting. This ensures that the bolt body fails before the part threads, since it is easier to replace bolts.

Any analysis has to be blended with experience and common sense. Analyzing thread stripping is a complex subject, and this analysis is only for the case where the threads have a lower yield strength than the bolt. Simple analysis has a place in the overall risk assessment. What is the consequence of a stud strip-out? Is it an orderly shutdown, or is a major leak with the potential for a catastrophic event? An analysis such as this provides additional quantitative input which can be used in making the final decision.

10.3.4   Case History: Analyzing a Spline Failure

The failure shown in Figure 10.5 involves a rear axle spline shaft, but similar failures have been analyzed in turbo machine gearboxes. Metallurgical evaluations help identify the various spline failure modes. Fretting and wear are usually associated with misalignment or lubrication issues. Fatigue failures can result from cyclic torques, such as occur from torsional vibration or bending fatigue from failed couplings. In fatigue, the cracks progress from each spline radius (Figure 10.6) until the shaft fractures through. There is usually no visible distortion, as in Figure 10.5.

Stress concentrations at the spline root won't affect gross yielding significantly but can cause fatigue cracks and depend on the material properties and surface treatment. Fatigue is not analyzed in this example since a twisted spline is the signature of a ductile bulk yielding of the shaft due to excessive torque. At the point where the spline is shown twisted (Figure 10.5), the spline was not supported torsionally by the hub. The question is: What torque does it take for such a permanent deformation, and where can it come from?

For this case the design torque delivered and the nominal shaft shearing stress at the bottom of the spline away from the root radius of the splines were

$$T = \frac{63,000(\text{hp})}{\text{rpm}} \quad \text{in.-lb}$$

$$S_s = \frac{16T}{\pi D^3_{\text{root}}} \quad \text{lb/in}^2$$

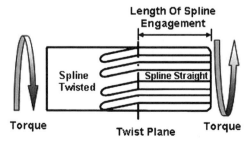

FIGURE 10.5   Twisted spline failure.

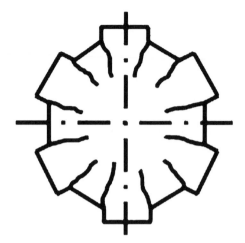

FIGURE 10.6   Spline fatigue cracks.

The following data are for full power operation, with all power transmitted through one axle: $D_{root} = 1.25$ in.; hp $= 100$, axle rpm $= 500$;

$$T = \frac{63,000(100)}{500} = 12,600 \text{ in.-lb}$$

$$S_{Sroot} = \frac{16(12,600)}{\pi(1.25^3)} = 32,860 \text{ lb/in}^2$$

The shear stress in yield of this steel is 70,000 lb/in^2. When the nominal stress calculated reaches this point, the shaft surface will be at yield but there will be no permanent shaft distortion since the bulk of the shaft will still be elastic. For permanent deformation, theoretically it will take at least $1\frac{1}{3}$ times the surface yield torque [45].

To yield across the entire cross section required the drive torque to have been higher than

$$\frac{70,000}{32,860}(1.33) = 2.8$$

What could cause such a torque overload? One scenario, which was the correct one, was that the vehicle got stuck in mud, both wheels were spinning and then one suddenly hit the pavement, stopping the vehicle almost instantaneously. The limited-slip differential gearing could fail or the spline could fail. In this case it was the spline. The owner's manual cautioned against such operation.

Even with power removed, the rotating masses were still trying to "wind up" the shaft. A 2.8 impact value can easily be exceeded by the drivetrain components suddenly being stopped. An example of sudden stoppage and the impact torque is given in Sections 2.13 and 10.3.11. For this case, under the most favorable transmission setting, the impact was calculated as 3.5.

Redesigning a spline requires that the next weak link in the system also be identified, as it could result in a more serious failure. The manufacturer may have designed the spline shaft as the weak link for a reason. Basic observations and calculations are useful in proposing cause scenarios, discounting others, and evaluating solutions. This is a valuable contribution to any troubleshooting and repair effort and helps prevent repeat failures.

10.3.5   Case History: Bending of Impeller Blades

The impeller blades were being bent on a new axial compressor used to draw process gas from a reactor. One additional blade was noted bent at each inspection. Compressors in similar service at other sites were not experiencing this problem. The investigating team observed that 1-in. polymer balls formed and passed through the upstream debris screen. Also these impeller blades were thinner than on other same-service machines. An analysis was requested to determine the effect of a polymer ball striking the blade tip.

The plan was to evaluate the compressors that were not bending any blades and see if they would bend with the same impact force. If they didn't, one solution would be to use blades of the same thickness. Why not just make the blades thicker and forget the analysis? The concern was that if the true cause were not known, more extensive damage could occur. Also, the thinner blades had a higher efficiency. The intricate geometry of the impeller is simplified in Figure 10.7 and illustrates the bent blade region. The polymer ball is shown rolling down the inlet pipe, where it is struck

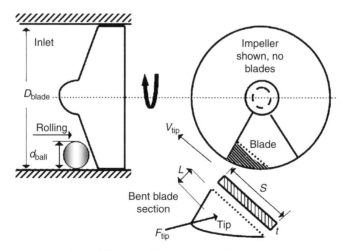

FIGURE 10.7   Impeller struck by product.

by one of the revolving blades. The force required to yield the blade permanently along line S where it was bent is

$$F_{bend} = \frac{zSt^2\sigma_{yield}}{6L} \qquad lb$$

Here z is a factor to yield the cross section through, and is equal to 1.5 [46].

The force due to the polymer ball being struck by the blade with all the energy absorbed by the ball distortion is

$$F_{impact} = V_{tip}\left(\frac{W_{ball}k_{ball}}{g}\right)^{1/2}$$

k_{ball} is the spring rate of the polymer ball and is determined by testing. For this analysis the pertinent data are $V_{tip} = 5655$ in./sec; $W_{ball} = 0.017$ lb; $S = 2$ in.; $L = 2$ in.; $t = 0.1$ in.; $k_{ball} = 313$ lb/in.; $\sigma_{yield} = 100,000$ lb/in^2; and $g = 386$ in./sec^2. The force required to bend the blade and the force available due to the ball impact are

$$F_{bend} = \frac{1.5(2)(0.1)^2(100,000)}{6(2)} = 250 \text{ lb}$$

$$F_{impact} = 5,655\left[\frac{0.017(313)}{386}\right]^{1/2} = 664 \text{ lb}$$

Since $F_{impact} > F_{bend}$, the blade will bend. Bent blades have not occurred on machines with 0.25-in.-thick blades instead of 0.1-in. blades. The force required to bend a 0.25-in.-thick blade is

$$F_{bend} = \frac{1.5(2)(0.25)^2(100,000)}{6(2)} = 1563 \text{ lb}$$

Here, the impact force is less than the force to bend, so a blade of this thickness should not bend. This explains why no other blade bending has been experienced. The blades bent one at a time, since debris only carried through the screen periodically during upsets. Since a smaller mesh screen would cause too much pressure drop, and it was impossible to stop product carryover, an impeller with a thicker blade geometry was installed, and the efficiency loss was accepted reluctantly.

10.3.6   Case History: Compressor Rod Failure

Troubleshooting technical problems can be much like investigating a mystery. This is why most engineers love a good failure, as long as it is not theirs. Facts have to be assembled, data have to be analyzed, and possible causes have to be verified. This case involves the wreck of a gas engine compressor. This is a 2000-hp unit operating at 300 rpm and is used in an ethylene refrigeration circuit.

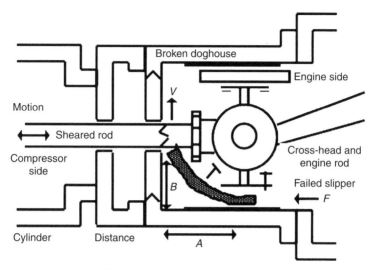

FIGURE 10.8 Compressor rod failure.

The wreck occurred on one of the two high-pressure cylinders. All that was left in the "doghouse" section was a pile of cast iron rubble. The 3-in.-diameter compressor rod was broken in a low-stress area away from the end threaded into the crosshead. The broken end of the rod was beaten over. The rod failed in direct shear. The operator said that he heard one loud bang. The wreck broke out the back end of the doghouse, as shown in Figure 10.8. This required over 1,000,000 lb of force. For the rod in tension to fail required over 1,000,000 lb, and for the rod to fail in direct shear required 500,000 lb. A force on the crankshaft of over 50,000 lb would have damaged the bearings, but there was no evidence of this.

There were bolt thread imprints on the broken part of the "slipper" holes, which is the cast iron bearing surface bolted to the crosshead. The bolts were shown to have failed in shear. Further review of the damage indicated an imprint of the compressor rod diameter on an end piece of the broken slipper. The question was: What caused the high loads, and why did the 3-in.-diameter shaft fail in shear?

From the evidence, let's test the hypothesis that the bolts came loose and the slipper moved back and forth affecting the hold-down bolts until they sheared, thus the imprints. The lower slipper wedged itself in the angled position shown. When the crosshead came forward, it struck the slipper, causing the loud bang. The force was not enough to damage the bearings (i.e., not over 50,000 lb). Using static equilibrium, the wedging shear force effect (V) of the slipper between the rod and crosshead can be estimated:

$$V = \frac{FB}{A} \quad \text{lb}$$

$$= \frac{50,000(16)}{2}$$

$$= 400,000 \text{ lb}$$

$$\text{shear stress } S_S = \frac{4}{3} \times \frac{V}{A_S} \text{(impact factor)} \qquad \text{lb/in}^2$$

$$= \frac{4}{3}\left(\frac{400,000}{0.785D^2}\right)(2)$$

$$= 150,000 \text{ lb/in}^2$$

A stress of 150,000 lb/in² was enough to shear the rod. The broken rod, which was still attached to the crosshead, could then strike the cast iron doghouse, resulting in a brittle failure of the doghouse while pulverizing the cast iron slipper. The solution was to dowel the slippers to the crosshead and increase the bolt load so that the dowels and the clamping friction took the horizontal load. Bolts were also found to be loose on the low-pressure slippers, which helped to verify the cause.

10.3.7   Case History: Seal Failure Due to Misalignment of an Agitator Shaft

Agitators are mixing devices used to blend products. Some products require a seal on the shaft, and the shaft must run true to prevent leakage. In this case history, a high-speed agitator had a shaft that was coupled together in sections. On startup the agitator shook uncomfortably with a high-frequency vibration. This case history examines what caused the problem and how it was solved.

Figure 10.9 shows the agitator mounted to the top of the vessel. A static runout reading was taken at the impeller and was measured as $e = 0.02$ in., due to tolerance problems with the shaft couplings. There were several possibilities for the vibration,

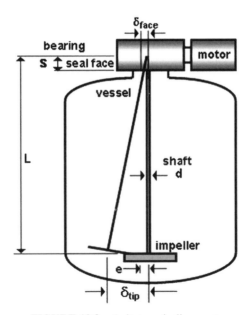

FIGURE 10.9   Agitator misalignment.

and two questions that needed to be answered were whether the runout was causing the vibration and if it were, whether it would cause the mechanical seal to leak. The following analysis was used to help answer these questions.

The unbalanced force due to an eccentricity of the mass and causing the vibration is

$$F = 28.4 \ We \left(\frac{\text{rpm}}{1000} \right)^2 = 35 \text{ lb}$$

where $W = 20$ lb., $e = 0.020$ in., and rpm $= 1750$. The deflection at the impeller shaft end due to this unbalanced force when assumed fixed at the bearing is

$$\delta_{\text{tip}} = \frac{FL^3}{3EI} = 0.29 \text{ in.}$$

where $E = 30 \times 10^6$ lb/in^2, $d = 3$ in., $I = \pi d^4/64 = 3.976$ in^4, and $L = 144$ in. The approximate runout at the seal face

$$\delta_{\text{face}} = \frac{s}{L} \delta_{\text{tip}}$$

$$= \frac{24}{144} (0.29) = 0.048 \text{ in.}$$

The vibration alone probably won't cause much of a problem other than loosening some bolts, but a seal operating at 1750 rpm and with a runout of 0.048 in. will probably leak. Generally, it is considered prudent to keep the static runout at less than 0.001 in. per foot of shaft length. The vibration and runout were reduced to acceptable levels by precise alignment and doweling of the shaft sections.

The solution seems obvious once the analysis was done, but at the time, under the pressures of a startup, there were several possible causes. Since this was a new installation, there could have been a critical speed, too flexible a mounting structure, gearbox, motor, or coupling, impeller problems, or the need for a steady bearing. In this case the least time consuming and least costly solution was tried first, which is usually a good approach when troubleshooting.

10.3.8   Case History: Gear Tooth Pitting Failure

Although gears can fail in a number of ways, one common degradation mechanism is pitting of the teeth. This can be due to excessive loading, and this is often because the driver has been uprated but no consideration was given to the rest of the system. Gearbox failures can be ugly. Gears are usually very badly pitted, with teeth broken out and bearings failed.

This case history is concerned with a new ship system. Pitting of the pinion was noted after only 2000 hours of service. The teeth did not break but were quite noisy. The following analysis was done to determine if the gears were susceptible to pitting at the operating loads. If they were just repairing the gearbox, that would not be

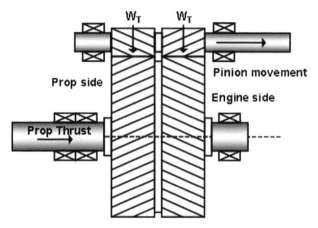

FIGURE 10.10   Gear system with pitting.

acceptable, as it would probably fail again in one year. On the other hand, if the analysis indicates that it should not fail, it would be necessary to find the cause elsewhere in the gearbox. In either case a quantitative conclusion could be reached if an analytical model were developed.

A failed gearbox on a ship at sea is not a happy situation, especially in a storm, as the ship cannot be positioned safe from danger. Since this was a fishing vessel, it was also possible for many tons of perishable cargo to be lost before the vessel could be towed to port. The model shown in Figure 10.10 was developed for the gear system. Things of importance are the loads on the gear face and how to determine if these loads are high enough to shorten the life by pitting. The bull gear did not experience any significant pitting since a gear tooth experiences a lower number of cycles than the pinion. It turns more slowly.

Reference [47] provides guidance on performing a gear pitting analysis, given here in a simplified form. The load on a gear mesh as a function of the input power and geometry is

$$W = \frac{126,000(\text{hp})}{d(\text{rpm})}$$

For this system, hp = 3000, rpm = 1000, pitch diameter $d = 11$ in., and

$$W = 34,364 \text{ lb}$$

Saying the load is split evenly between the pinion mesh:

$$W_T = 17,200 \text{ lb}$$

The K factor is

$$K = \frac{W_T}{Fd} \frac{M_G + 1}{M_G}$$

The face width $F = 8$ in., $M_G = 200$-tooth bull gear/40-tooth pinion $= 5$, and

$$K = 234.5$$

The contact stress S_C will be calculated, as it is used to determine the pitting life:

$$S_c = C_K (KC_d)^{1/2}$$

The de-rating factor

$$C_d = \frac{C_a C_m C_s}{C_v}$$

The application factor $C_a = 1.8$ for engines, the load distribution factor is $C_m = 1.5$, the size factor $C_s = 1.2$, the dynamic factor $C_v = 0.8$, the geometry factor $C_K = 4336$, and

$$C_d = \frac{1.8(1.5)(1.2)}{0.8} = 4.05$$

The surface contact stress on the tooth is

$$S_c = 4336[234.5(4.05)]^{1/2}$$
$$= 133,625 \, \text{lb/in}^2$$

This stress is related to the pitting life by the curve shown in Figure 10.11 for case-hardened gear teeth. In the figure, the cycle life is 4×10^9 cycles. This ship sees 2.5×10^8 cycles per year on a pinion tooth. So this gear should not show pitting for

$$\frac{4 \times 10^9 \text{ cycles}}{2.5 \times 10^8 \text{ cycles per year}} = 16 \text{ years}$$

This is an acceptable design life for a fishing vessel. Unfortunately, it never made it and failed by pitting in less than one year. What could have been the cause?

In the analysis it was assumed that the load was evenly split between the helix and with each accepting half of the load W. What would happen if the load were not shared, that is, if the pinion did not "float" to balance the load? The pinion was designed to float since the propeller thrust load could move the bull gear a small amount during full-ahead and full-astern conditions. The load would not be split evenly if the outer races of the pinion bearings were not free to slide and center the pinion as the bull gear moved axially. To review the worst case, the value of K

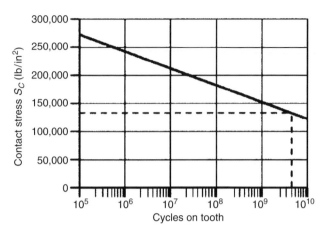

FIGURE 10.11   Gear tooth pitting life.

can be doubled, as it contains the gear load. Doubling K means either that the load on the pinion had doubled or that only half of the face width is in contact. It could also mean that a little of both is occurring, as would occur with a partial sharing of the load and deflection of the tooth under load.

$$S_c = 4336[234.5(2)(4.05)]^{1/2}$$
$$= 189,000 \, \text{lb/in}^2$$

Now the design life is only 4×10^7 cycles or a couple of months, which is about the time it took for the noise and pitting to be observed. Other calculations were performed on the gearbox to evaluate its adequacy, but modifying the pinion outer race bearing fit so that it was free to slide and split the load solved the pitting problem without a gear redesign.

Although the analysis seems simple enough, the author can still remember the turmoil that occurred. There were many other factors that could have caused the gears to pit. Distortion of the pinion, twisting of the gearbox, errors in the gear geometry or gear case machining, or torsional vibration of the system were some of the scenarios that surfaced. This is where analytical modeling is extremely useful. It allows the analyst quantitatively to verify or eliminate possible causes. Eliminating potential causes can be just as important as finding the most likely cause in such an analysis.

10.3.9   Case History: Impact Load Effect on a Large Gearbox Bearing

A gearbox for a 14,000-hp extruder was having periodic bearing failures. A metallurgical report on the failed bearing mentioned that "stress butterflies" were present and that this type of transformation was an indicator of a high-stress condition. Sometimes the bearings cracked the race, and other times the races spalled. An analysis of the loads indicated that a twofold torque could be experienced under certain startup conditions, but even with this load, the high stresses could not be explained.

A "successful failure" occurred when a bearing failure produced additional information. In this case, evidence was observed on one side of the bearing which correlated with the roller positions. It was also reported that operators had tried repeatedly to start the machine with the extruder full of cold material, which locked up the output shafts. The bearing has an axial clearance δ and it was suspected that during hard startups the gears and shaft mass suddenly moved through the clearance and caused an impact of the rollers on the bearing thrust half. The investigation team speculated that this resulted in a "hammerlike blow." This impact could have been large enough to damage the race and thus greatly shorten the operating life.

The following analysis was performed to verify or disprove this theory. If it was verified, a soft-start system would be installed, operating procedures would be modified, and several previous failures would be explained. Figure 10.12 shows the input torque being applied to the gear. The axial component of the load on the helical gear is the load P. The resulting force on the roller elements is shown as Q. Notice that the other set of rollers is not shown since they are unloaded in the thrust direction and the other end is free to float. The free-body diagram (FBD) of the system is shown in Figure 10.12, which greatly simplifies the problem. The spring constant k (lb/in.) is for the bearing and structure and is the effect of the bearing dilating from the internal loading and the gear case stiffness.

From this model the reaction load F on the rollers due to the clearance δ and load P can be developed. The impact force of a mass at some velocity against a spring, assuming that all of the kinetic energy is absorbed by the spring, is

$$F = V\left(\frac{W}{g} \times k\right)^{1/2} \qquad \text{lb}$$

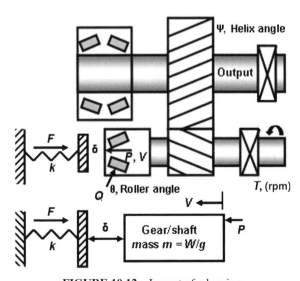

FIGURE 10.12   Impact of a bearing.

For constant acceleration (a) starting from zero, the unknown is the velocity:

$$V = (2a\delta)^{1/2}$$

For a force acting on a mass,

$$P = ma = \frac{W}{g}a$$

$$a = \frac{Pg}{W}$$

Substituting and simplifying yields

$$V = \left(\frac{2Pg\delta}{W}\right)^{1/2}$$

$$F = \left(\frac{2Pg\delta}{W}\right)^{1/2}\left(\frac{W}{g} \times k\right)^{1/2}$$

$$F = (2P\delta k)^{1/2} \qquad \text{lb}$$

Now we have an equation that links the impact force of the moving shaft/gear mass to the input torque, suddenly applied, through a clearance. One thing also obvious is that there is a wedging action taking place as the load F pushes on the rollers and results in load Q. This is the load that will be used to determine the bearing stress.

For a wedging action with N even number of rollers all in contact, at an angle θ (deg) the load per roller is

$$Q = \frac{F}{N \sin \theta}$$

$$Q = \frac{(2P\delta k)^{1/2}}{N \sin \theta} \qquad \text{lb}$$

Now P (lb) is the axial component of the input torque T (in.-lb), rpm, gear pitch radius R, helix angle ψ (deg), J a startup torque impact factor, and L_{gear} the load on the gear due to the input torque T:

$$\text{hp} = \frac{T(\text{rpm})}{63,000}$$

$$T = \frac{63,000(\text{hp})}{\text{rpm}}$$

$$L_{gear} = \frac{JT}{R}$$

$$P = L_{gear} \tan \psi$$

Putting this all together, the equations to determine the impact force Q on a roller are

$$Q = \frac{(2P\delta k)^{1/2}}{N \sin \theta}$$

where

$$P = \frac{J(63,000)(\text{hp}) \tan \psi}{R(\text{rpm})}$$

The Hertzian contact stress is highest for a roller on the inner race:

$$S_c = 3240 \times \left[\frac{(Q/z_{\text{roller length}})(D_{\text{race}} - D_{\text{roller}})}{D_{\text{race}} D_{\text{roller}}} \right]^{1/2} \quad \text{lb/in}^2$$

This value can be compared with the basic static load rating stress, where permanent deformation occurs and life is shortened. For a single roller this stress is around 580,000 lb/in². For this system,

$$
\begin{array}{ll}
\text{hp} = 14,000 & k = 30,000,000 \text{ lb/in.} \\
\text{rpm} = 1200 & \theta = 10° \\
\delta = 0.02 \text{ in.} & \psi = 20° \\
z_{\text{roller length}} = 3 \text{ in.} & R = 6 \text{ in.} \\
N = 21 & D_{\text{race}} = 16 \text{ in.} \\
& D_{\text{roller}} = 2 \text{ in.}
\end{array}
$$

Table 10.1 shows the effect of three cases on the bearing contact stress margin. The margin from permanent deformation of 1.3 with a hard start is much too close. Indications were that the impact load was shared by only half of the rollers.

The soft-start margin is better and has not resulted in similar failures on machines that had this option. It still appears high relative to the normal operating stress. This can be deceptive since the soft start will gently take up the 0.020-in. clearance, so that impact does not occur. This being the case, the soft-start system will probably have a larger margin than shown. All of this makes it quite probable that impact was the problem and that a soft-start system and better startup procedures could eliminate these types of failures.

TABLE 10.1   Bearing Impact Stress Cases

Condition	Torque Multiplier, J	Clearance, δ	Contact Stress, S_c	Margin $= 580,000/S_c$
Hard start	4.0	0.020	441,000	1.3
Soft start	0.6	0.020	274,000	2.1
Normal operating stress	1	0.0007	134,000	4.3

Obviously, reducing the clearance would help significantly. Loading the gear set with spring washers so that there is no clearance but while allowing axial growth is one possible solution. In this case it was not possible without considerable redesign of the gearbox, and the soft-start option was preferred.

10.3.10   Case History: Motor Shaft Failure

In most industries, about 95% of failed equipment problems are remedied by replacing the defective parts, and no analysis is performed. Under the pressure of production, this allows the operation to be back in business quickly. When a costly repeat failure occurs soon after the first, the failed part is often sent to an outside materials laboratory for determination of the failure mode.

For example, consider the case of a failed motor shaft connected to a pulley and V-belt that drive a small mixer (Figure 10.13). In this case the laboratory report provided some valuable information. It stated that the failure mode was a rotating bending fatigue failure in the sharp radius of a snap ring groove. Often, the symptom is treated instead of the cause, as it was in this case. The snap ring groove was provided with a generous radius to reduce the stress concentration and the equipment was returned to service. Unfortunately, a similar costly failure occurred again within one week of intermittent operation.

The following simple analysis was used to identify the true cause and to prevent a repeat failure. The force due to a properly tightened V-belt from Section 2.11 is

$$F_V = \frac{303,000(\text{hp})}{D_P(\text{rpm})}$$

The bending moment at the snap ring groove is

$$M = F_V L$$

The bending stress in the snap ring groove is

$$\sigma_B = \frac{kMc}{I}$$

In this familiar equation, $k = 2.5$, $c = D/2$, and $I = \pi D^4/64$.

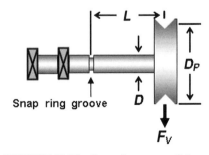

FIGURE 10.13   Snap ring groove failure.

For the motor assembly, hp = 150, rpm = 1750, D = 2.5 in., D_P = 36 in., and L = 12 in. The cyclic bending stress as the shaft rotates is calculated as

$$\sigma_B = 14,109 \, lb/in^2 \text{ tension to compression}$$

The rotating bending endurance limit for this material is 30,000 lb/in^2.

Since there is no mean bending load, the factor of safety is simply

$$FS = \frac{30,000}{14,109} = 2.13$$

According to this analysis, even with this sharp radius snap ring groove, the shaft should not have failed in fatigue. But it did! So either the laboratory report is in error or something has occurred that wasn't considered in the analysis. One possibility is that the torsional load hasn't been included and there may have been mixer impact problems. Further investigation indicated that this was not the case. Another possibility is that the bending stress was higher than calculated and is what caused the rapid failure. The question, then, is what could have caused higher stress.

The belt load equation includes a factor to account for a belt tightened correctly. This load and thus the stress can easily be multiplied several times by excessive tightening. Shafts have been known to yield statically when jack screws are used to tighten the belt incorrectly. What occurred in this case is that a new machinist had not been trained on the correct belt-tightening procedure. Some training and instructions on a placard on the belt guard solved the problem. When something has failed and the analysis says that it shouldn't have, determining why this is so can usually lead to the true cause.

10.3.11 Case History: In-Flight Aircraft Crankshaft Failure

One thing great about engineering is that so many interesting problems can be modeled that can apply elsewhere. Although this case is an aircraft crankshaft failure, it could have been a bearing seizure in a hydrocarbon processing compressor. The math is the same but the experience is not quite as personal.

A single engine aircraft experienced a crankshaft failure in flight at 9000 ft altitude at night in the clouds. This is every pilot's worst nightmare. The propeller stayed attached to the aircraft and was rotating independently, in the first main bearing. The engine was ticking away merrily with everything shaking violently. The pilot circled down through the clouds, and to his amazement there was an airport directly below. Talk about luck.

This analysis was performed to determine why the crankshaft failed. Since the two parts of the crankshaft had broken and destroyed the fracture surface, a metallurgical analysis provided only a few clues. Fatigue-related cracking and some inclusions due to a "dirty" steel were observed. The pilot mentioned that a propeller stoppage had occurred several years earlier, when the aircraft was taxiing on a grass runway. With the engine idling at 250 rpm, the nose wheel went into a hole and the propeller blade hit the turf, stopping the engine.

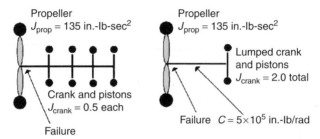

FIGURE 10.14   Mass–elastic diagram of aircraft engine.

To determine if this stoppage could have caused a fatigue crack, the following model was developed. The "fishbone" sketch of Figure 10.14 represents a mass elastic diagram of the aircraft engine with the mass inertia of the cylinders shown as J_{crank} and the propeller as J_{prop} (in.-lb-sec^2); C represents the shaft stiffness, (in.-lb/rad). When the propeller is suddenly stopped, the inertia of the crankshaft and the pistons want to keep rotating and twist the crankshaft journal, web, and pin at the failure point. This occurs until all of the kinetic energy of the rotating mass is converted to potential energy associated with twisting the journal. This is a very simplistic model, as the energy is dissipated in other ways, too.

The engine develops 150 hp at 2500 rpm and the full-load torque T at the failure point is

$$T = \frac{63,000(\text{hp})}{\text{rpm}} = 3780 \text{ in.-lb}$$

The torsional moment M due to the sudden stoppage from 250 rpm can be determined by equating the kinetic energy to the potential energy and was discussed in Section 2.13:

$$M = \frac{\pi(\text{rpm})}{30} (JC)^{1/2}$$
$$= \frac{\pi(250)}{30} [2(5 \times 10^5)]^{1/2} = 26,200 \text{ in.-lb}$$

This impact torsional moment is $26,200/3780 = 6.9$ times higher than the normal operating load torque. An analysis by the manufacturer has the maximum design stress under normal operating loads as 21,000 lb/in^2 at the failed area, and the crankshaft should be able to withstand 120,000 lb/in^2. The stresses in the crank journal/pin web overlap area are complex but approximately directly related to the torque. With a 6.9 multiplier, or a $6.9(21,000) = 145,000$ lb/in^2 stress, there is a high probability that a crack could have developed at the time of stoppage and eventually propagated through the crankshaft by the normal cyclic operating stresses. Machines remember abuse, as shown definitively in Section 2.28.

Engine teardown, inspection, and a new "clean" steel crankshaft made using the VAR (vacuum arc re-melt) process, was installed. With the most probable causes determined and mitigated with the improved steel and lack of further propeller stoppages, no further problems occurred.

10.3.12 Case History: Pitting Failure Due to a Poorly Distributed Bearing Load

Pitting on half of a double-row roller bearing was noted in a five-year-old gear unit during a planned downtime inspection. Misalignment of the shaft was suspect due to the wear pattern. The outer bearing was a large single-row roller bearing and was not experiencing pitting. To verify or discount this claim, the system was modeled with the rollers as a group of springs in parallel. The reason for the springs goes back to the basics of analytical modeling: to make sure that you include in your model parameters that can be modified.

Since it was suspected that shaft deflection could cause misalignment, it was reasoned that bearing and structure stiffness could also be of significance. It might not be possible to change bearing or structure stiffness, but it would be possible to change the bearing type if there was a problem with alignment. By modeling the combined rollers and gearbox stiffness as springs, the load distribution at each spring due to misalignment could be determined (Figure 10.15). The loading, speed, shaft, and bearing sizes and gearbox and bearing stiffness are fixed, so not much can be done with those parameters. However, if the analysis indicates a poor load distribution, and that distribution can be shown to result in a short bearing life, the question becomes how to improve the load distribution.

The analysis now has a purpose, and that is to determine the load distribution. This was not just a spark of intuition by the model builder. Those involved with this machine suspected that there was poor load distribution by observing the bearing failures and the results from the materials laboratory. There was already some discussion about installing a spherical bearing to handle the misalignment and to better distribute the load.

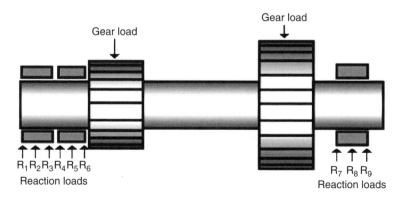

FIGURE 10.15 Bearing load distribution.

TABLE 10.2 Reaction Loads on Bearings in Pounds

R_1	R_2	R_3	R_4	R_5	R_6	R_7	R_8	R_9
3,000	8,000	14,000	17,000	25,000	32,000	32,000	24,000	18,000
0	0	0	0	0	100,000	74,000	0	0

The following analysis was used to describe the concern to the bearing and gear unit manufacturers. The purpose was to better understand the problem and to try and persuade the manufacturer to change the bearing design. The bearing was known to be much more flexible than the gear housing, and the manufacturer provided a bearing stiffness of 4×10^7 lb/in. This was divided equally between the rollers, and a structural program was used to determine the reaction loads. This is what is summarized in the first row of Table 10.2. The second row illustrates an unlikely extreme: that the bearing and support structure were rigid. Notice how with no flexibility the load concentrates at the corner of the roller and no distribution of the load occurs. This would occur because there is no capability to adjust for misalignment. This usually does not occur in machinery because there is always some flexibility to accommodate misalignment.

Reviewing the first line of data in the table, notice how the load is distributed along the rollers. It appears that most of the load is on the second roller. As a rough calculation we can use the loading on the second roller (i.e., $17,000 + 25,000 + 32,000$) and reduce the double roller C rating by $\frac{1}{2}$. This can then be compared with the full C rating, which assumes even loading between the rollers:

$$B_{10} = \frac{1.93}{\text{rpm}} \left(\frac{C}{P} \right)^{10/3}$$

For the case where the designer assumed that the load was evenly distributed across the two rollers,

$$B_{10} = \frac{1.93}{800} \left(\frac{750,000}{100,000} \right)^{10/3}$$

$$= 2 \text{ years} \qquad \text{about 14 years of average life}$$

For the case where the load was only over one roller,

$$B_{10} = \frac{1.93}{800} \left(\frac{750,000}{2/74,000} \right)^{10/3}$$

$$= 0.54 \text{ year} \qquad \text{about 3.87 years of average life}$$

Notice that the B_{10} life has been multiplied by 7 to obtain the average life, which is really what needs to be considered when troubleshooting. Remember that the B_{10} life is the statistical life 90% of a bearing group will survive, and the median life is the life 50% will survive and is about five times the B_{10} life. The average is about seven times the B_{10} life.

Several estimates have been made in the analysis, such as the bearing stiffness. If the bearing were less flexible, the loads on this roller would be higher and the life would be less. In any case, the analysis does not predict a long life, and the 3.8-year life predicted is confirmed from the pitting noted. Discussions with the manufacturer on the benefits of going to a spherical bearing design to extend the overhaul interval would be worthwhile.

The gear manufacturer and bearing manufacturer performed a detailed analysis and implemented the modification. It was important that the manufacturer agreed with the modification and made the necessary design change. In this way the design responsibility remained with the manufacturer and not the user.

10.3.13   Case History: Failure of a Preloaded Fan Bearing

Not all failures are high-cost, lost-production failures. Some are just—well—embarrassing. This is a case where a fan system (i.e., a motor driving a fan through a belt system) was failing every three to four weeks, and the unit had gone through six sets of bearings. The plant had purchased the unit from a used equipment supplier, and no drawings or operating instructions were available. The manufacturer had gone out of business many years ago. Every time the bearings failed, they were replaced with the same type of pillow block ball bearings. The fan was located remotely and automatic greasing had been used, which was functional. On the last failure, a failure analysis was requested by the site technician. A review of the failed bearing indicated that it had experienced high thrust loading. Although ball bearings can take some thrust load, they are designed primarily for radial loads.

From Figure 10.16 it is obvious that the major loading is going to be from the radial belt load and also from the radial weight load of the overhung fan. The thrust loading on the bearings should be negligible. The only way a sizable thrust load could occur is if the shaft got hotter than the base and the shaft grew in length. This

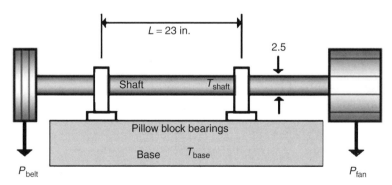

FIGURE 10.16   Loading of a fan bearings.

would not be a problem with a typical design of this type. In such a design, one bearing is fixed, usually on the pulley side, and the other bearing is allowed to float. This allows the shaft to grow and not load up the bearings. When the manufacturer of the pillow block ball bearings was contacted, it was stated that the ball bearings being used were both "fixed" and each would allow only 0.001 in. of axial play. The technician said the fan shaft T_{shaft} was about 30°F hotter than the base T_{base}. The following analysis was done to determine if this could be the cause of the high thrust loads.

From calculations the belt load was determined to be 700 lb radial load and the fan weight is 200 lb radial. At these loads the bearings will provide a B_{10} life of 10 years, which is more than adequate and certainly won't cause a failure in three weeks or show an axial load on the bearing race. Now consider the axial expansion of the shaft length between the bearings:

$$\delta_{shaft} = \alpha L(T_{shaft} - T_{ambient})$$
$$\delta_{base} = \alpha L(T_{base} - T_{ambient})$$

The growth of the shaft assuming it is hotter than the base is

$$\delta = \alpha L(T_{shaft} - T_{ambient}) - \alpha L(T_{base} - T_{ambient})$$
$$\delta = \alpha L(T_{shaft} - T_{base})$$

For this case, $\alpha = 6.6 \times 10^{-6}$ in./in./°F, $L = 23$ in., $T_{shaft} - T_{base} = 30$°F, and

$$\delta = 0.0046 \text{ in.}$$

Divided equally this means that each bearing needs to be able to accommodate 0.0023 in. of shaft expansion before it will start to preload. But the manufacturer has told us that each bearing can accommodate only 0.001 in. This means that each bearing will be preloaded by $0.001 - 0.0023 = -0.0013$ in. This is not good for these types of bearings. An idea of the preload can be approximated by the following equation, where P_{axial} is the preload in the shaft due to this -0.0013-in. preload and A_{shaft} is the cross-sectional area of the shaft of diameter 2.5 in. This was shown in Section 2.6.

$$\delta_{preload} = \frac{P_{axial}L}{A_{shaft}E}$$

Rearranging and solving for P_{axial} gives us

$$P_{axial} = \frac{\delta_{preload}A_{shaft}E}{L}$$
$$= \frac{0.0026(0.785)(2.5)^2(30 \times 10^6)}{23}$$
$$= 16,600 \text{ lb axial load per bearing}$$

The loads are probably lower than this, but this was enough information to change the bearing design on the fan side to a "floating" bearing. The original design probably had a fixed–floating bearing design. However, somewhere along the way a fixed–fixed design was installed. The same bearings kept being purchased, as there was no records of what was in the original design. These are not easy problems to solve. However, when the bearings show high loads where there should not be high loads, this was valuable information for the analysis. In this case thrust loading was evident by the ball tracking on the bearing races.

10.3.14 Case History: Separating Loads in an Extruder

One very complex machine is an extruder used to process polymers. The final product comprises plastic pellets sent to customers to mold into plastic products. There are many other types of extruders in the rubber- and food-processing industries. They are quite large, with some over 14,000 hp. When one of these units fails, it can shut down production for many days.

Cases on gear failures associated with these types of machines have been discussed, but wear can be a major concern on big machines. In Section 2.34 it was shown that the load had to be known to apply the wear equation. In an extruder, the separating load that is pushing the screw against the barrel needs to be known. This type of information is extremely important when a new machine, which may be the largest ever built, is in the design stage. Since the customer has information on the wear of smaller machines, it will be shown that this type of information can be used in troubleshooting other machines. A case history in wear will help show this.

For now, all that is of interest is determining what load is pushing the screws against the barrel. Figure 10.17 is a model of high-pressure area loading. The pressure buildup in the intermeshing region results in lateral forces that try to push the screws apart. The pressure in these high-pressure areas is very complex and act on areas of the screw. The resulting forces act on the screw, which push them against the extruder barrel. No attempt is made here to determine the exact direction of these forces. The simplification is made that they are directed perpendicular to the barrel surface.

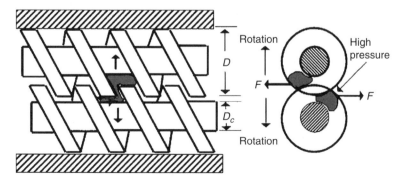

FIGURE 10.17   Extruder high-pressure loading area.

The power into each shaft is dissipated in heat generation due to friction and shearing and pressure development. The following analysis is based on treating each screw separately, so that half of the motor horsepower is applied to each shaft. A power balance on a shaft can be represented as the power in being equal to the power dissipated:

$$hp_{screw} = hp_{heat} + hp_{pressure}$$

The power dissipated in heat, using the specific heat equation, is

$$hp_{heat} = \frac{WC\Delta T}{2544}$$

From the geometry of the extruder screw, and saying that the torque in the high-pressure areas results in separating forces F,

$$hp_{pressure} = F\frac{D_{eff}}{2}\frac{rpm}{63,000}$$

where $D_{eff} = (D + D_c)/4$. This can then be solved for the separating force F:

$$F = \frac{126,000}{D_{eff}(rpm)}\left(hp_{screw} - \frac{WC\Delta T}{2544}\right) \quad lb$$

With some actual operating data on an extruder: $hp_{screw} = 684$; $rpm = 200$; $D_{eff} = 9$ in.; $C = 0.5$ btu/lb-°F; $\Delta T = 350°F$; and $W = 9000$ lb/hr,

$$F = 4542 \text{ lb force acting on each screw in the high-pressure area}$$

An idea of how the input horsepower is dissipated is

$$\frac{hp_{heat}}{hp_{screw}} \approx 0.90$$

$$\frac{hp_{pressure}}{hp_{screw}} \approx 0.10$$

This is an approximate indication which shows that 90% of the power goes into raising the product's temperature, and only 10% into developing the pressure.

This is a very approximate method for determining the loads in an extruder, and it is not easy to verify that they are correct. However, this was done to evaluate in-service extruders that had a large database of wear history. By having an approximation of the internal extruder forces, the wear could be determined using the methods shown in a wear case history. The wear measured was used as an indirect method to determine if the force calculations were reasonable. Certainly, a direct method to measure these forces would have been preferable.

10.3.15   Case History: Containment of an Impeller

Containment of rotating failed parts is a real safety concern. We saw in Section 6.2 how the gas explosion inside a cast iron manifold resulted in flying fragments. Luckily, no one was injured. This case history examines a turbocharger impeller that came apart. The fragments were contained in the housing, but the question came up of how safe this was and whether the high-speed impeller could burst through the housing if such a failure happened again. The manufacturer had many of these units in service. Conducting an overspeed burst test might not tell much since it would not explain how thick the housing would have to be to contain the parts. The engineer did this rough calculation to get some idea of what it would take to break through the housing.

The fragments left after the failure were broken into many small pieces. This is probably because they were ground up. It is known that flywheels, fans, and impellers usually fail at the bore, resulting in three or more pie-shaped sectors. For this analysis it will be assumed that the impeller failed in three sections. More sectors would each have less energy associated with them and cause less damage. From high-speed photographs of bursting impellers, flywheels, and fans it was shown that the pieces move out radially at the bore and also in the direction of rotation. This is shown clearly in Figure 10.18.

For this problem the interest is if the sector, which is shown moving at velocity V, will break through a housing of thickness t, of which only part of the circumference is shown in the figure. All sectors move out equally tangentially but for this sketch only one is shown. The velocity of the sector tangentially from the centrifugal effect is shown as

$$V = \frac{2\pi \, \mathrm{rpm} \, R_B}{60} \qquad \text{in./sec}$$

The velocity at distance s, which is the clearance of the impeller to the housing from the time it fails is constant.

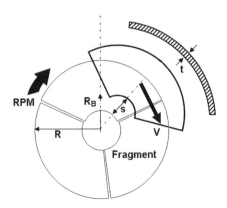

FIGURE 10.18   Bursting impeller.

It will be assumed that all of the kinetic energy equals the potential energy, which means that the sector sticks to the housing, which is not true but is conservative:

$$\frac{\frac{1}{2}W}{g} V^2 = P\delta$$

Solving for δ yields

$$\delta = \frac{WV^2}{2gP}$$

Now δ is how much the sector weight W penetrates the housing, and P is the force developed to do this:

It is time for another assumption based on the metal punching process, as when a piece of steel is punched out by a die in a press. The force required to do this is the area in shear times the shear strength of the material:

$$P = \tau A_{shear}$$

Assume that the area in shear has perimeter L. For a rectangle, $L = 2(\text{length} + \text{width})$, and for a circle, $L = \pi D$ (in.). This is as if the sector circumference contacts the housing.

The distance δ is the penetration depth:

$$P = \tau(L\delta)$$

Putting this in the equation and solving for δ, we have

$$\delta = \frac{WV^2}{2g\tau L\delta}$$

$$\delta = V\left(\frac{W}{2g\tau L}\right)^{1/2}$$

Substituting the expression for V and simplifying yields

$$\delta = \frac{\pi}{30}(\text{rpm})\, R_B\left(\frac{W}{2L\tau g}\right)^{1/2}$$

As long as the thickness of the housing, t (in.), is thicker than the penetration depth δ, the part should not "punch out":

$$t \text{ in.} \gg \delta \text{ in.}$$

Knowing the weight W (lb), the shear strength of the housing τ (lb/in²), the clearance distance between the impeller and the housing s (in.), and the sector perimeter L (in.), the penetration depth δ (in.) can be determined. Some data:

$$
\begin{aligned}
W &= 3\,\text{lb} & t &= 0.25 \text{ in.} \\
\text{rpm} &= 10{,}000 & s &= 1 \text{ in.} \\
\tau &= 50{,}000\,\text{lb/in}^2 & L &= 12 \text{ in.} \\
R_B &= 3 \text{ in.} & g &= 386 \text{ in/sec}^2
\end{aligned}
$$

$$
\begin{aligned}
\delta &= \frac{\pi}{30}\ \text{rpm}\ R_B \left(\frac{W}{2L\tau g} \right)^{1/2} \\
&= \frac{\pi}{30}\,(10{,}000)\,3 \left[\frac{(3)}{(2)12(50{,}000)(386)} \right]^{1/2} \\
&= 0.25 \text{ in.}
\end{aligned}
$$

From this analysis one could be concerned since the housing is 0.25 in. thick and the equation calculates the penetration to be 0.25 in. or concludes that it will punch through.

The housing hasn't failed even though the calculation shows that it should. This indicates that the equation is in error and that one of the following may have occurred:

- The impeller broke up into smaller sectors than the shape assumed, thus not having enough energy to break through. However, next time it may not break up.
- Not all of the kinetic energy goes into punching out the case; some of it goes into deforming the housing or pulverizing the impeller. However, next time it may not.
- The punch-out mode of failure may not be appropriate.

In any case, it is reasonable to gather other sources of containment information in order to determine if the model is reasonable.

Containment data are not easy to obtain, and the author had to rely on data available on racing flywheel explosions and jet engine fan blade explosions, some of which are shown in Table 10.3. With these limited data, it appears that although the model is in error, the order of magnitude is about correct. So there is cause for concern that loss of containment is possible, meaning that a sector of the impeller could punch through the housing if it failed again.

Having a mathematical model reveals some important information:

$$
\delta = \frac{\pi}{30}\ \text{rpm}\ R_B \left(\frac{W}{2L\tau g} \right)^{1/2}
$$

TABLE 10.3   Comparison of Failure Data with Containment Equation

Item	Weight (lb/sector)	Tangential Velocity (in./sec)	L (in.)	s (in.)	τ (lb/in²)	Calc. δ Model (in.)	Test δ (in.)
Racing burst flywheel	7	2200	20	0.5	50,000	0.21 in. *Says OK*	Contained 0.38-in. steel bell housing
Racing burst flywheel	7	2200	20	0.5	20,000	0.33 in. *Says fail*	Through 0.38-in. aluminum bell housing
This case history	3	3200	12	1	50,000	0.25 in. *Says fail*	Contained in 0.25-in. steel housing
Helicopter turbine rotor	3.6	4500	38	0.5	50,000	0.23 in. *Says fail*	Through 0.14-in. 304L sheet

- The potential for punching through is directly proportional to the speed, so limits overspeeds.
- A secondary high-shear-strength steel shield τ around the housing $\frac{1}{4}$ in. thick should provide additional containment, for added protection.
- A lighter impeller would help if the unit is ever redesigned.

Although this seems pretty basic, recall that before the model was constructed, no one had any idea if the present $\frac{1}{4}$-in. case was adequate or what the margin was from loss of containment.

The containment error margin from loss of containment is defined as

$$\text{containment margin} = \frac{\text{actual penetration}}{\text{penetration } \delta \text{ (in.) calculated}}$$

At least now the containment margin is more in the range 0.6 to 1.8 rather than 0.1 to 10, 100, or even 1000. Sometimes, such order-of-magnitude calculations are all the specialist really wants the analysis to provide, to validate the need for additional costly testing.

10.4   FAILURES CAUSED BY WEAR

10.4.1   Case History: Examining the Wear of Extruder Screws

In this case history, an analytical model was developed for predicting wear life, in years, for polymer twin extruder screws and barrels. The model allows the owner to incorporate historical extruder wear information to better define the wear process. This is a rather detailed model that took considerable effort to build and evaluate and has proven to be a tool that provides significant insight into the extruder wear process using the wear equation. The basics of the model are quite simple, as are the mathematics.

With new extruders some of the largest ever made, and with much existing extruding equipment being run at high throughput to meet market demand, it is important to be able to anticipate and address any potential extruder barrel and screw wear life concerns before they happen. Predicting the potential for unacceptable wear life early on existing units allows for a planned and managed shutdown at a time that will minimize lost opportunities and maintenance costs.

The type of wear analyzed is an increase in clearance between the screw flights and the barrel (δ). Excessive wear can result in screw breakage, product contamination or nonuniformity, product degradation, or a reduction in pumping capacity. Wear tests of hard-surfacing materials under laboratory conditions provide a valuable guide for extending extruder screw and barrel life. Unless the laboratory tests have simulated the actual operating conditions in the extruder, such tests cannot quantitatively determine the wear life when in operation. An evaluation within an operating extruder would provide the most accurate results; however, it would be prohibitively costly and time consuming, and control of the many variables is difficult if not impossible in a production environment. Scale-up of small laboratory extruders results to those expected on full-sized units must be made with care.

Users of extruders routinely measure wear of screws and barrels during downtimes and use this type of information when planning for screw or barrel replacement, when consulting with the extruder manufacturer, and when reviewing new designs or troubleshooting existing designs. This model was developed to help the user organize and better utilize this historical database.

The analysis method is based on the wear model of Section 2.34, and considers the forces, developed in Section 10.3.14, which contribute to the wear loads on the screw and barrel: the shaft weight and the effect of the product in the high-pressure region of the extruder. The model inputs the loads and rub length acting on a material surface. An experimental wear coefficient K, determined from historical extruder wear performance, is used to define conditions at the wear surface. The constant thus determined will indicate the type of boundary condition present, that is, the range of conditions between a nonlubricated surface to a hydrodynamic film interface. The output represents the time for a given amount of wear depth to occur, or in the case of this model, the amount of radial wear on the flight or the barrel. The loading is based on considering the screw to be supported by the flights and the separating loads to be a percentage of the input torque.

This model was developed to help answer some of the following historical questions, which are asked routinely by extruder users:

- The hard surfacing is harder but thinner (nitrided). Will the wear life of the screw be increased or decreased?
- For scale-up of an extruder outside the present operating envelope, with no historical data for this extruder size, can the wear life be predicted based on historical wear information?

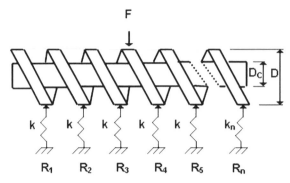

FIGURE 10.19   Extruder screw reaction loads.

- This is a strange type of wear pattern in the barrel and on the screw. What could have caused it, and what should be done?
- Will all the startups and shutdowns result in an unacceptable increase in the wear, and at what interval should a planned inspection be scheduled?

Determination of the Reaction Loads Causing Wear   The extruder screw is represented as a beam on multiple flexible supports k (Figure 10.19). Only one of the screws is shown. These flexible supports represent the extruder barrel stiffness. With the load distribution known, this problem can be solved using available software, such as structural design programs. The reaction loads can then be used to represent the load on a flight.

Separating Forces on Twin-Screw Extruders   Since the unbalanced load acting on the extruder screw must be known, the appropriate applied torque that produces separating forces must be determined. This is probably the most difficult unknown to determine. Strain-gauging extruder barrels to determine loads would be an accurate method to use. In operating units this is generally not possible, so analytical approximations by modeling were done. This is shown in Section 10.3.14. Knowing these loads is the key in determining wear. For now we can say that only about 10% of the input torque into the screw generates pressure in extruders and that most of the rest is expended as heat. This results in the high-pressure region in an extruder, where 10% of the input torque can be converted into separating forces acting on each screw. From the separating load results,

$$F = \frac{9450(\text{hp}_{\text{in}} \text{ per screw})}{D_{\text{eff}}(\text{rpm})}$$

where $D_{\text{eff}} = (D + D_C)/4$. In this equation it is assumed that the screws receive 75% of the motor horsepower [48]. This separating force can then be used to determine the reaction loads on the flights by using multiple support beam theory (i.e., the screw is supported by the flights). For the rigid-support case used here, the flight reaction loads R_3 and R_4 in the high-pressure region were each approximately $0.5F$. D is the screw diameter (in.) over the flight tips, D_C is the channel diameter, as shown in Figure 10.19, and rpm is the shaft speed.

The wear process in an extruder is complex, and the most important wear mechanisms in extrusion are adhesive, abrasive, and corrosive wear. No comprehensive wear model of the process has yet been developed, and it is far beyond the scope of this work to develop one. The model developed in Section 2.34 is used to analyze the wear. It considers essentially two rubbing surfaces, with boundary lubrication between them and normal load acting on the surface.

By converting the geometry of the surfaces into that of an extruder's flight, linear velocities to angular, and assuming that the surface rubs during the cycle, the time for the extruder surface to wear an amount δ in a given time t can be determined. In equation form this is represented as

$$t = \frac{(\text{BHN})\delta}{1162KD(\text{rpm})\sigma} \quad \text{years}$$

$$\frac{\delta}{t} = \frac{1162KD(\text{rpm})\sigma}{\text{BHN}} \quad \text{in./yr}$$

where D is in inches and σ is in lb/in². By considering the depth of hardness and the allowable depth of wear in the base metal, the wear rate for each can be determined, assuming that the other variables stay constant. The total wear life can then be determined.

Thus, with the typical geometry, operating conditions, material properties, and wear coefficient K, the life of an extruder screw can be determined. The wear coefficient K, which depends on the lubrication conditions between the barrel and the flight, can be determined using historical data. With the radial wear of the screw in the high-pressure region measured, and the time for this amount of wear to occur recorded, the value of K can be calculated by rearranging the equation and solving for K. The significance of K alone is useful. When K is less than 10^{-7}, good surface boundary lubrication conditions can be expected, and thus low wear [23, p. 443]. All wear would probably be due to startups, shutdowns, and process upsets when metal-to-metal contact is possible. With K greater than 10^{-4}, an unlubricated condition can be expected to exist, and with high loads or poor wear-resistant surfaces, high wear rates could be expected. Thus, just calculating K can reveal the condition between the flight and the barrel. For many polymer machines considered, K is in the 10^{-8} range and supports the opinion that the screw is supported by the polymer [49].

Analyzing the Results   Figure 10.20 indicates the effect of the hardness of the wear surface. This is for an extruder that has a cobalt flight tip hard-face layer 0.125 in. thick with a hardness value of 380 BHN (Rc 40). The flight base material has a value of 200 BHN (Rb 100). The wear rate in the hard-face material is calculated as

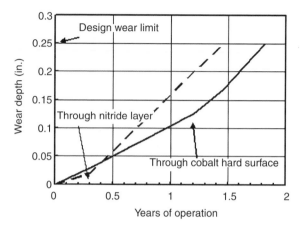

FIGURE 10.20   Extruder hard surface effect.

0.11 in./yr and is through this hard layer in 1.2 years, and in the softer material the wear rate accelerates to 0.20 in./yr. This is approximately proportional to the hardness. Since this extruder can tolerate up to 0.25 in. of radial wear, the life is calculated as being 1.8 years. The K value based on the life calculations is 2×10^{-7}, which indicates that the polymer is providing good support but is an order of magnitude greater than for that experienced on polymer extruders, where the K value is 2×10^{-8}. Cracked hard facing due to poor application and dilution contributes to the lower value on this particular machine [50]. Figure 10.20 also illustrates an example of a surface that is very hard but also thin. When the flight is nitrided to a depth of 0.02 in., instead of applying the cobalt hard facing, the wear rate is only 0.06 in./yr through the nitride layer. However, it progresses through this thin layer in 0.3 year and spends the rest of its life with accelerated wear in the softer base material, shortening the overall life.

Table 10.4 indicates the real value of such a simplified model. It can show the effect of changing several variables at once. This is what usually occurs on new, larger installations, when all the conditions have been changed. This is extremely difficult and costly to evaluate in the field, since it would require control of the operating variables and periodic screw removal to measure the wear.

In this example an estimation of the wear life of extruder B is required when information is available on the present operation of extruder A. In Table 10.4 the wear limit life was know for extruder A, and the K value was calculated based on this. For this reason the variability of the wear coefficient K is not a factor, as it is assumed to remain constant for evaluating extruder B. It is shown only to illustrate the excellent lubrication provided by the polymer at the screw–barrel interface.

In this example, the hardness and thickness were not changed by the manufacturer, and the flight width was increased to reduce the contact pressure. Even by doing this, the effect of the higher pressure, screw speed, and modified geometry increased the life by 25%. Reducing the hardness or thickness of the hard-surface

TABLE 10.4 Effect of Changing Extruder Variables on Wear

Parameter	Extruder A	Extruder B
Rpm	200	220
Peak pressure (lb/in²)	1000	1600
Lead (in.)	8	8
Flight width (in.)	0.5	1.0
Barrel diameter (in.)	8	8
Hard-surface BHN	380	380
Hard-surface thickness (in.)	0.125	0.125
Base metal BHN	200	200
K (unitless)	2×10^{-8}	2×10^{-8} based on A
Wear limit (in.)	0.125	0.125
Life to wear limit (years)	3.4	4.25

treatment would have further reduced the screw life. The new life was acceptable for this extruder and product, so no further modifications were required.

Use of The Model as a Troubleshooting Tool   Reviewing a new extruder design on the potential for excessive wear is an important consideration. However, it often happens that an in-service extruder that has had a historically acceptable life suddenly has its life reduced dramatically. The question then is: Why is the life shorter, and what can be done to correct the problem? When the model is showing that this extruder should have a long life but in actuality it is not, something has changed.

Using this type of problem-solving approach has resolved problems involving excessive growth and interference with the barrel, and powder packing. Although these parameters are not analyzed by the model, further examination of the separating loads or pressures may be required. For example, if a sensitivity analysis indicates that loads much higher than those calculated are required for the amount of wear that has occurred, the possibility of powder packing or thermal interference should be reviewed with the extruder manufacturer.

The model can also be used to eliminate some of the troublesome variables. For example, in one case on a large polymer extruder, a slight initial wear problem was thought to be due to the weight of the screw. Much energy was expended, together with heated discussions with the manufacturer of the equipment. Analyzing this case indicated that the weight of the screw was not significant relative to the life, so this variable was eliminated from further consideration. Later, it was determined that the cause was an injection line installed at the incorrect location. The problem was quickly resolved.

In a final case, an extruder that had a good life history and was predicted by the model to have adequate life, was experiencing a life reduced by more than 80%. The cause was a large number of dry startups and shutdowns, due to problems in making a new product. Basically, the screw was starting up frequently and running for a considerable length of time with no product in the barrel, thus greatly accelerating the wear. When no information is available on the number of years of life required to produce a given amount of screw wear on similar extruders, a wear coefficient value will have to be estimated.

Even controlled wear testing can yield K value variations of more than 2 from controlled test to controlled test [23, p. 478]. Although this may be of concern to engineers, one must take into account the wide range of wear coefficients encountered, from $K = 10^{-4}$ to 10^{-9}, a range of 100,000. As mentioned, reducing the uncertainty in predicting the amount of wear likely to be encountered from a factor of 100,000 to a factor of 4 or even 10 is clearly an accomplishment. When [23], K is determined from life data it does not represent a controlled test, in that the load was also estimated. Thus, the K value does not really indicate the boundary condition alone, but includes the uncertainty of the load calculated over the contact area (i.e., σ).

The uncertainty can be reduced by using the model to examine the relative influence on wear performance of changes in operating conditions or materials (i.e., comparing wear in two different extruders, with the wear life of one being known). Conditions that have a large degree of variability, such as the wear coefficient, can be considered to be constant between the two extruders, as in the Table 10.4 example, and only the change in wear life due to an increase or decrease in the other variables must be evaluated, thus obtaining a relative picture. This picture is far closer to being correct than is the absolute value.

10.4.2   Case History: Wear of a Spline Clutch

This case involves the use of one of the wear equations developed in Section 2.34. We analyze the life of a large clutch that was used to engage and disengage a 11,200-hp motor to a gearbox. It contained gear teeth splines to allow for clutch disk alignment and contained rubber bushings that were supposed to provide flexibility. The alignment of the motor to the gearbox through this coupling was quite stringent for both the radial (offset) and axial (angular) alignments. There were three of these splined disks, each with 100 teeth, to take the input load.

The gearbox became misaligned due to thermal bowing of the extruder barrels, which is described in Section 10.5.1. This misalignment had worn grooves 0.050 in. deep in the splines, which eventually locked up and cracked the input shaft to the 50-ton gearbox and also generated additional damage within the gearbox. It was a costly repair with considerable production impact.

The problem is that after the clutch and gearbox were repaired, reinstalled, and back in operation, the gearbox moved again and became misaligned with the motor. Each revolution of the shaft caused the disk gear to wipe back and forth on the spline gear, wearing away material. The amount of misalignment in the operating system was known since by geometry, the coupling face axial misalignment could be determined from the displaced position of the gearbox. The question asked was whether with the amount of misalignment indicated, could the gearbox be operated until the next planned downtime, which was in two years.

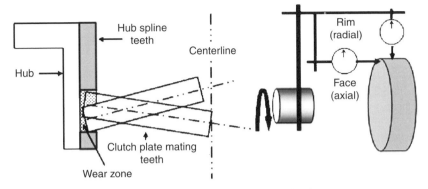

FIGURE 10.21   Wear of a spline clutch.

This is the real benefit of analytical modeling. Data on wear of regular cou-plings cannot be used since geared couplings are lubricated and have hardened teeth. When the grease in a geared coupling turns to sludge, separates out, or dries up, the geared coupling quickly fails. This coupling is unlubricated and has splines with a Brinell hardness of 150, which is quite soft. With an analytical model, the goal is to determine how long it will take to wear to the 0.050 in. depth under the present operating conditions. Actually, since it is known that lockup occurs at the 0.050 in. depth, it would be prudent to stay well away from this amount of wear.

Figure 10.21 shows only one of the three spline clutch disks. The misalignment wiping action is also shown, along with how this misalignment would be determined with dial indicators, one stationary and the other with the arrow rotated. The time it takes for the coupling to move through the rubbing distance δ_{axial}, which is the total indicator reading (TIR) that a dial indicator would see during one revolution on a coupling face reading at the spline is $t = 1/\text{rpm}$ min/cycle. The velocity of the gear face rubbing this over the full cycle (zero to maximum and back to zero) is

$$V = \frac{2\delta_{axial}}{t}$$

$$= 2\delta_{axial}(\text{rpm}) \qquad \text{in./min}$$

When radial misalignment is also present, the velocity can be determined by vector addition:

$$V = 2(\text{rpm})[(\delta_{axial})^2 + (\delta_{radial})^2]^{1/2}$$

The load on a tooth is

$$F = \frac{63,000(\text{hp})/\text{rpm}}{RN_{teeth}N_{disks}}$$

The stress on the face of a tooth is

$$\sigma = \frac{F}{wh} \qquad \text{lb/in}^2$$

In Section 2.34 the various equations for wear were developed. The equation that is used here is

$$t = \frac{(\text{BHN})\delta}{370K\sigma V} \qquad \text{years}$$

For this case history,

BHN = 150	h = tooth height = 0.47 in.	δ_{axial} = 0.020 in.
hp = 11,200	rpm = 1200	N_{disks} = 3
w = tooth width = 0.73 in.	R = 14 in.	N_{teeth} = 100

$$V = 2(0.020)(1200) = 48 \text{ in./min}$$

$$F_{\text{tooth}} = \frac{63,000(11,200)}{3(100)(14)(1200)} = 140 \text{ lb}$$

$$\sigma = \frac{140}{0.73(0.47)} = 408 \text{ lb/in}^2$$

Recall that K is a wear coefficient that is available in textbooks or from actual field data. In this example, field data are available since a preliminary examination of the gear tooth wear showed 0.005 in. of wear after only one month of operation, with 0.020 in. axial and no radial misalignment. Initially, a paste lubricant was used on the teeth, which probably results in the much lower K value shown. Book values say that for dry metal-to-metal contact, $K = 1 \times 10^{-4}$ to 1×10^{-5}.

$$K = \frac{(\text{BHN})\delta}{370t\sigma V}$$

$$= \frac{150(0.005)}{370(\frac{1}{12})(408)(48)}$$

$$= 1.2 \times 10^{-6} \qquad \text{unitless}$$

$$t = \frac{(\text{BHN})\delta}{370K\sigma V}$$

$$= \frac{150(0.050)}{370(1.2 \times 10^{-6})(408)(48)}$$

$$= 0.86 \text{ year or about 10 months}$$

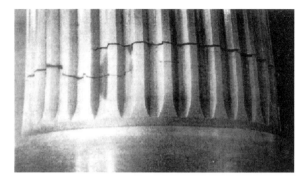

FIGURE 10.22   Input shaft to gearbox.

With the current misalignment of 0.020 in. axial, it is highly unlikely that the spline will last in operation for as long as two years. In less than a year, lockup of the splines could be expected to reoccur.

The realization that another broken input shaft could occur, along with a severe production impact, forced the decision for a planned downtime. The condition that caused the misalignment was addressed, and the motor was aligned to the gearbox to within 0.002 in. radially and axially. A 0.002-in. radial and axial misalignment should result in a coupling life of six years, which is adequate, as the machine has a planned downtime every two years. The coupling teeth can be lubricated at that interval. The cracked spline shown in Figure 10.22 was the input shaft into the gearbox. It illustrates the heavy loads that can result because of a shaft lockup. These are not the peripheral clutch splines analyzed but are the clutch center drive splines.

10.5   FAILURES CAUSED BY THERMAL LOADS

10.5.1   Case History: Thermal Distortions Move a 50-Ton Gearbox

This case history considers the uneven heating of a long extruder barrel. Figure 10.23 shows an extruder motor, gearbox, and barrels. The extruder is used for processing polymer, and the heat generated by the process is removed by cooling passages in the barrels. When the passages get fouled, side-to-side temperature differentials occur. The gearbox weighs 50 tons and is bolted to a base plate. It moved even with the bolts tightened adequately. When a gearbox moves, unpleasant things happen. Couplings wear and lock up, input shafts crack, and bearings start to fail.

This simple model was used to predict when the temperature differentials were excessive and the passages had to be cleaned. Consider the simple plan view of the gearbox and the extruder in Figure 10.24. Uneven heating causes the barrel to bow and imparts a thermal moment C_{applied} onto the gearbox. The problem is to determine if this thermal moment is greater than the resisting moment. M_{resist} caused by the bolting and weight frictional load between the gearbox and the base plate. If C_{applied} is greater than M_{resist}, the gearbox will move and misalignment of the coupling can be expected.

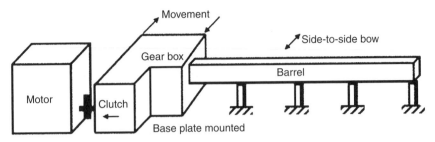

FIGURE 10.23   Extruder system.

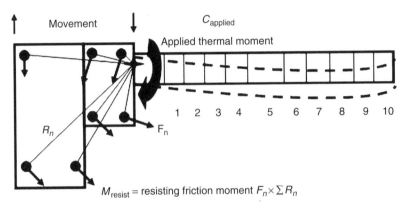

FIGURE 10.24   Forces on system due to thermal moment.

To determine the magnitude of $C_{applied}$, a simple three-dimensional finite-element model was used. The actual measured temperature profile was put on the model to simulate side-to-side temperature differentials. Table 10.5 shows the results of one such analysis. Verification of the model was done by checking the calculated deflection with laser-measured deflections. $C_{applied}$ of 267,000 ft-lb is the twisting effect that the gearbox weight and bolt holddown force have to resist.

The resisting moment M_{resist} (ft-lb) is

$$M_{resist} = FR_1 + FR_2 + \cdots + FR_7 = F \, \Sigma R_n$$

TABLE 10.5   Forces and Moments on Extruder System[a]

Location	1	2	3	4	5	6	7	8	9	10	11
ΔT (°F)	Ref.	62	97	39	72	56	28	21	85	78	Ref.
δ (mils)	0	12.8	28.5	36.5	48	54.7	55.7	58.2	60.2	41.5	0

[a]$C_{applied}$ = 267,000 ft-lb.

Inserting the radial distances R_n in the resisting forces F yields

$$M_{\text{resist}} = F(1.7 + 5 + 11.1 + 5 + 12.9 + 6.4 + 1.7) = 43.8F \qquad \text{ft-lb}$$

F is the effect of frictional resistance or the coefficient of friction μ times the normal force to the surface. The normal force to the surface is due to the gearbox weight, which is 50 tons or 100,000 lb, and the bolt holddown preload force, which is 50,000 lb/bolt.

$$F = \left(\frac{100,000}{7 \text{ bolts}} + \frac{50,000}{\text{bolt}} \right) \mu = 64,300\mu$$

$\mu = 0.05$ since there is oil between the gearbox and the sole plate:

$$F = 3220 \text{ lb and } M_{\text{resist}} = 43.8(3220) = 141,000 \text{ ft-lb}$$

Since $C_{\text{applied}} > M_{\text{resist}}$, the thermal moment due to the side-to-side temperature difference will move the gearbox and cause alignment problems.

From the model the effect of cleaning the passages and lowering the temperature differential and the effect of removing the oil from the base plate can be evaluated. Dowelling of the gearbox was considered, but it was feared that the restraint might distort the gearbox and cause internal problems with the gears and bearings. The solution was to remove the oil and clean the fouled passages so that ΔT was less than 30°F side to side.

10.5.2 Case History: Thermally Bowed Shaft

There have been many cases of bowed shafts caused by inappropriate warm-up procedures on steam turbines and compressors. Fouling can also result in seal rubs, which can cause thermal bows. All are noted by an increase in rotor vibration levels. Experienced engineers who have addressed this problem previously sometimes use cycling procedures to eliminate the bow. Making these tough decisions online can be an extremely cost-effective approach and are not for the faint-hearted. A wrong diagnosis of the cause can result in disastrous results. When site personnel have never experienced such vibration and decide to try the cycling procedure mentioned, a very different scenario can evolve, with a major wreck as the final outcome. For this reason a simplistic discussion of what causes a thermal bow may help to explain the seriousness of the problem.

Consider the shaft in Figure 10.25. When a zone on one side of the shaft is hotter (i.e., dark shade) than the other, that side tries to elongate. It can't, because it is restricted by the cold-side material. The only way it can get longer is to bow in the direction of the hot side. If the shaft were heated through (i.e., light shade), it would just grow axially and there would be no bow. That is why running a machine with a slight thermal bow and cycling it can sometimes cause it to "straighten itself," since the temperature equalizes and ΔT disappears. Figure 10.26 illustrates the small

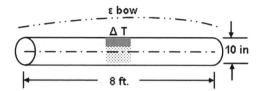

FIGURE 10.25   Simple shaft heated on one side.

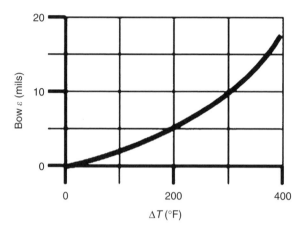

FIGURE 10.26   Effect of ΔT on shaft bow ε.

temperature difference needed on one side of this shaft to bow (ε) in the direction of the highest temperature. With clearances (δ) in the range 10 to 20 mils, you can see why an excessive thermal bow can be damaging.

Once the shaft bows and touches, which can be from an imbalance, gravity sag, or uneven heating, the same spot will continue to touch as it rotates. The frictional heat will increase the contact zone temperature. The more the temperature rises, the more the bow, the harder it rubs, and the higher the heat. Thus, a self-generating failure mechanism will develop. This is shown in Figure 10.27, with the shaft shown at four different positions as it rotates. An illustration of this would be to hold a slightly bowed wire between your fingers and roll it. Only one side of the bow will rub on an imaginary seal shell. Notice that with ample clearance δ, the rub will not occur, but imbalance and vibration will be present, due to the bow eccentricity ε of the rotating rotor mass.

Running with high vibration, assuming a bow, and hoping that the temperatures will equalize out or that the vibrations will "shake" things back into place is dangerous. It may be something else, such as missing blade imbalance, rotor instabilities, or a mechanically bent shaft. Without the necessary precautions, the bow can quickly magnify and cause a severe wreck. Taking a downtime, removing the rotor, replacing it with a spare, and having a large bowed shaft straightened by a reputable shop using heating and hanging techniques generally constitutes a much lower risk approach.

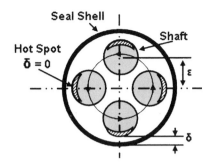

FIGURE 10.27   Rotating bowed shaft in seal.

10.5.3   Case History: Steam Turbine Diaphragm Failure

A 40,000-hp process turbine brought down a plant's operation when one of the 12 diaphragms rubbed and caused extensive damage. This resulted in a barrage of questions that required immediate answers.

Steam turbine rubs are not uncommon, but major wrecks are. Possible causes include:

- Bowing of the shaft due to sag or uneven heatup
- Loss of a blade or foreign-object damage
- Overpressurization
- Fouling
- Thrust bearing bad or cold clearances set incorrectly

None of these causes were found to pertain in this case. A pressure load four times that of the normal design pressure would have been necessary to cause the type of failure observed. Management wanted a credible failure scenario that could explain such a pressure increase before putting the machine back online.

One theory was that the flow was choked at the stage where the diaphragm rubbed. *Choked* means that a shock wave developed and limited flow, thus allowing pressure buildup on the upstream side of the diaphragm. Normal turbine flow pressure could not have caused this.

Data indicated that a volume of liquid water was admitted to the turbine inadvertently. Since the location of entry eliminated a water slugging damage problem, it was hypothesized that the extra water turned to steam instantaneously and increased the flow. This caused a choked flow condition to occur.

A simple analysis modeled the turbine as several simple volumes and orifices. The diaphragm vane passages are shown as vane volumes in Figure 10.28. Steam occupies the volumes in the turbine, and the orifices are the nozzle areas the steam flows through in the diaphragms. A volume of water, shown as the cylindrical bucket, is assumed converted to steam instantaneously.

From steam tables [29] at 35 lb/in^2 and 281°F, which are the pressure and temperature in the volume being considered, the specific weight of water (V_{water}) is

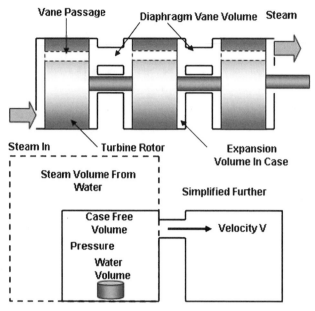

FIGURE 10.28   Steam turbine volume model.

0.0173 ft³/lb and that of steam (V_{steam}) is 8.5 ft³/lb. This means that a volume of water will expand 491 times its volume when converted to steam. In Figure 10.28 this is shown as the dashed line "steam volume from water."

From the steam tables, a simplification is that V_{steam} varies inversely with the pressure as shown, and V_{water} doesn't change much in the range of interest. Pressures are in lb/ft² absolute, and the approximate relationship for this is

$$V_{steam} = 8.5 \, \frac{7200}{P_{final}} \qquad \text{ft}^3\text{/lb}$$

$$V_{water} = 0.018 \qquad \text{ft}^3\text{/lb}$$

Since the volume is constrained by the turbine case, it cannot expand to this free volume but must be assumed squeezed into the casing volume. In doing this, the pressure must increase.

Using the perfect gas law and assuming that the temperature remains constant because the pressure rise is instantaneous gives us

$$P_{final} = \frac{V_{initial}}{V_{final}} P_{initial}$$

$$V_{initial} = \frac{V_{steam}}{V_{water}} W_{water}$$

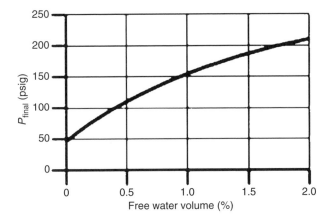

FIGURE 10.29   Water converted to steam pressure.

W_{water} is the actual weight of water converted to a volume:

$$W_{water} = 0.1334 \times \text{Gallons ft}^3$$

$$V_{final} = V_{case}$$

$$P_{final} = \frac{V_{steam} W_{water}}{V_{water} V_{case} P_{final}}$$

Making the necessary substitutions, solving for P_{final}, and simplifying yields

$$P_{final} = 673 \left(\text{gallons} \times \frac{P_{initial}}{V_{case}} \right)^{1/2} \qquad \text{lb/ft}^2 \text{ abs.}$$

A sensitivity analysis can be performed using the equation and P_{final} as the pressure. Figure 10.29 shows that it does not take much water converted instantaneously to steam in a confined volume to result in high pressures. In this case, water at 2% of the volume flashing to steam instantaneously raises the pressure to 220 psig.

There are problems with this analysis. The principal one is: How did the water in the fictional case get so hot as to flash into so much steam! If heat was added to the system continually, as in a fired boiler, it would be easy to understand; but it was not. The temperature the water reached in this case was not known.

Although the model didn't solve the problem, it was useful in understanding several scenarios that might occur. The final solution was to modify the startup procedures so that the problem couldn't reoccur.

10.5.4   Case History: Screw Compressor Rotor Rub

There are times when wear is suspected but cannot be confirmed without disassembling the machine. In this case history, this was experienced on a process screw compressor with two years of service. The efficiency and capacity of the compressor had dropped, resulting in operating losses. A process upset had occurred and the discharge temperature of the compressor had increased from 300°F to 500°F for a period of 2 minutes until it was brought under control. The question was if this could have caused the rotors to rub, which would have resulted in an increase in the clearance and a loss of efficiency.

The obvious thing to do would be to disassemble the compressor and inspect the rotors. However, there are many things other than a rotor rub that could have caused the loss of efficiency. To shut down the compressor would shut down the plant, resulting in loss of the production run. The question asked by the operations manager was: "Can we continue to run for a week to finish the production run?" If this could be done, there would be time to disassemble the compressor, inspect it, and replace the rotors if necessary, with no loss in production. A quick analysis was performed to see if a rotor rub could have occurred in such a short period of time. With a positive answer and future control of the temperature, the probability of being able to run for a week longer was high.

Figure 10.30 shows the simplified female and male lobes of the rotor system. When assembled, the cold running clearance δ is 0.003 in. Under normal conditions, the case and rotors heat up evenly and the operating clearance, even during hot-running conditions, is 0.003 in. according to the manufacturer. This had been verified, since no rubbing had been observed during previous inspections of the rotor.

For this analysis, assume that during a short period of time, the rotor lobe outlined, which is thin and of less mass (w) than that of the total rotor, will experience the temperature change much more quickly than will the total rotor mass. The lobe is engulfed in the high-temperature gas stream T_{inf}. This will cause the thin intermeshing lobes to "grow," from the increase in temperature, by

$$\delta_{hot} = \alpha L\Delta T \qquad \text{in.}$$
$$= 6.6 \times 10^{-6}(2.5)(500 - 300)$$
$$= 0.0033 \text{ in.}$$

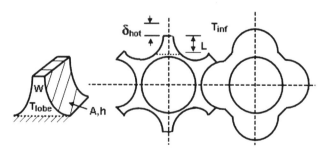

FIGURE 10.30 Rotary screw compressor model.

Since the cold assembly clearance was 0.003 in. and the running clearance is less, the possibility of interference (i.e., $0.003 - 0.0033 = -0.0003$ in.) could result in rubbing.

The question is whether the rotor lobe metal temperature could have increased from 300°F to 500°F in only 2 minutes. To determine this, the lumped capacitance method [51] is used to determine if the mass of the lobe could increase from 300°F to 500°F in 2 minutes. If it could, interference was possible. The simplified equation to determine the temperature T_{lobe} of lobe mass w after t hours is

$$T_{lobe} = \varepsilon^{-zt}(T_0 - T_{inf}) + T_{inf}$$

where $z = hA/Cw$ and $\varepsilon = 2.718$. In the equation above, the variables are

h = surface convection heat transfer coefficient (Btu/hr-ft²-°F)

C = specific heat of lobe (Btu/lb-°F)

t = time to reach T_{lobe} (hr)

A = surface area h (ft²)

w = lobe mass to be heated (lb)

T_{inf} = gas temperature lobe immersed in (°F)

T_0 = temperature of lobe at time $t = 0$ (°F)

For this case, $h = 150$, $C = 0.12$, $A = 0.3$, $w = 8$, $T_{inf} = 500$, $T_0 = 300$, and $t = 0.033$. Solving the equation gives us

$$T_{lobe} = 457°F$$

This shows that in 2 minutes it is possible for the lobe to reach almost 500°F, and it is quite possible that a rub has occurred. The decision was to continue operating for a week while carefully monitoring the temperature so that the temperature did not exceed 350°F.

On later disassembly of the machine it was noted that the discharge end of the rotor had rubbed, and the rotor was replaced. Faster temperature control, along with liquid injection, was implemented, which controlled temperature excursions. In addition, the cold assembly clearance of the new rotor was increased from 0.003 in. to 0.005 in. as an added precaution. This resulted in a slight loss in efficiency, which was tolerated for the increase in reliability. After seven years of operation, no further rubs had occurred.

10.5.5 Case History: Hidden Load in a Three-Bearing Machine

Production demands can stretch equipment to their design limits. Uprates should be well analyzed, as loads and temperatures can increase and possibly reduce the life of

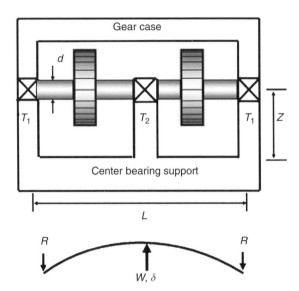

FIGURE 10.31   Three-bearing shaft support.

couplings, bearings, gearing, and shafting. In this case history, modifications increased the gearbox oil temperature, which resulted in additional "hidden loads" and repeated bearing failures.

The internal clearance of ball or roller bearings is the clearance between the rolling elements and the race. Adequate clearance is required to keep a bearing from being excessively preloaded during installation or shaft thermal growth. It is also critical when a shaft is supported by three bearings. If sufficient misalignment occurs at the center bearing, potentially destructive loading can be produced.

The simplified gearbox shown in Figure 10.31 was actually complex in design, with several intermediate shafts and gears transmitting high horsepower to a piece of processing equipment. It had been in satisfactory service for many years, and the motor size had recently been increased. Within a year of operation, the center bearing failed. An analysis indicated that the gear loads on the bearings were not excessive. The bearing was replaced, but failed again a year later.

It is important to understand any possible source of hidden loads on the center bearing. This is because a bearing's life is sensitive to load. For example, doubling the reaction load on a roller bearing can reduce its design life by 90%. When one recognizes the sensitivity of a third bearing on alignment and loading, the question of thermal growth of the center support needs to be considered. Modifications allowed the center bearing support to be 50°F higher than the end bearing supports. The question was if this could produce additional loading that needed to be added to the gear reaction loads.

The deflection δ of a shaft with simple support conditions at each end bearing and with an intermediate load W is also shown in Figure 10.31. The center load deflection is

$$\delta = \frac{WL^3}{48EI}$$

$$I = \frac{\pi d^4}{64}$$

The thermal growth of the center support is

$$\delta = \alpha\, z\, \Delta T$$

Since the shaft deflection and support growth δ are equal, the two deflection equations are set equal to each other and solved for the bearing center load W:

$$W = \frac{48EI\alpha z\, \Delta T}{L^3}$$

For the gearbox being considered, $E = 30 \times 10^6$ lb/in^2, $d = 6$ in., $\alpha = 6.6 \times 10^{-6}$ in./ in.-°F, $z = 20$ in., $L = 30$ in., $\Delta T = T_2 - T_1 = 50$°F, and $I = 63.4$ in^4, so

$$W = 22{,}390\,\text{lb}$$

Thus, a significant additional load can be produced if adequate internal clearance is not specified for the center bearing. Load sharing of the bearings is an important issue. Too much internal clearance and the center bearing will become unloaded and the end bearings will carry the total reaction load. Too little and the center bearing will take most of the load.

A new center bearing was ordered with increased internal clearance, as the end bearings had a higher load capacity. This reduced the center bearing loading considerably, which permitted operation until a long-term solution could be implemented. This analysis allowed an engineer to quantify what other experienced persons were speculating. Those difficult decisions as to whether equipment can continue to operate carry less risk and fewer sleepless nights when backed by a screening analysis that helps in understanding the cause.

10.6   MISCELLANEOUS FAILURES

10.6.1   Case History: Crack Growth in a Rotor

Crack growth calculations using fracture mechanics are discussed in Section 11.7 and the example used here represents the case history. Several large driers used to dry products were being installed in new plants. Similar units with less horsepower

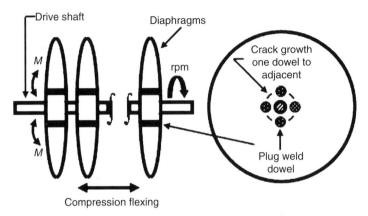

FIGURE 10.32   Drier rotor crack growth model.

were developing cracks. The rotors were such that they were like diaphragms, and with each rotation they flexed in and out with a resulting cyclic bending stress. A three-dimensional finite-element study, which was the most complex part of this analysis, indicated that the nominal stress fluctuation was 6 ksi. The geometry was such that the disks were welded with plug-welded spacers which had an initial crack length of 0.5 in., and this is where the maximum bending stress occurred. When the growth of the cracks reached a length of 2 in., it was possible for the cracks to join about the circumference and break the disk loose. This would cause the drier rotor to shift forward and result in a major wreck as the rotors struck the case. Management wanted to know if the new machines could be operated for two years, until a planned downtime, or if they had to be sent back to the factory to be repaired. Figure 10.32 is a cross section of the rotor assembly, which weighed several tons.

As shown in Section 11.7, the cyclic stress of 6 ksi was calculated as 2.19×10^6 cycles for the crack to grow from 0.5 in. to 2 in. Since the rotor rotates at 10 rpm, this is 14,400 cycles/day, and the time it would take the crack to grow to a 2-in length (i.e., $2a = 2$) is

$$\frac{2.19 \times 10^6 \text{ cycles}}{14,400 \text{ cycles/day}} = 152 \text{ days}$$

This is only five months, and since the load couldn't be reduced, the decision was that the drier could not operate for two years and had to be modified to lower the cyclic stresses.

Periodic inspection of the unit produced the points shown as crosses on Figure 10.33, which were then compared with the life calculation. The agreement was rather astounding, considering all the assumptions that were made. Linearizing the crack growth, that is, drawing a straight line and looking at the slope, indicates the growth rate:

0.5 in. per 300,000 cycles (21 days)

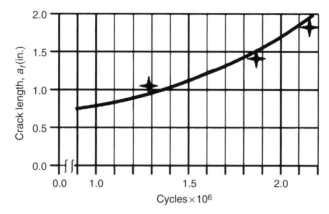

FIGURE 10.33   Crack grown in rotor diaphragm.

Some idea of the redesign required could be made as a check. The stress in a plate with a moment is proportional to the square of the thickness. Increasing the highest-stressed diaphragm plates from $\frac{1}{4}$-in. thickness to $\frac{3}{8}$-in. thickness would reduce the predominant bending stress to

$$(6 \text{ ksi})\left(\frac{0.25}{0.375}\right)^2 = 2.7 \text{ ksi}$$

This cyclic stress magnitude was enough to keep any cracks from propagating and was the design change incorporated, together with better welding procedures.

10.6.2   Case History: Structural Failure Due to Misalignment

Although most of the case histories have concerned primarily operating or mainte-nance elements, sometimes the problems concern poor design details. The failure in this case occurred six months after installation of a large drier-type vibrating con-veyor used to transport and dry an elastomer-based product. The equipment was large, over 30 ft long and 8 ft high, constructed of welded I-beams, plates, and a stainless steel sheet metal pan that contained the product. It transported the product using a vibrating pan which moved in an arc-type motion by means of stabilizer bars or struts.

After six months of operation, extensive and severe cracking was noted in the structural steel components, localized near a connecting strut, also called a *stabi-lizer*. It was not known when the cracking had begun. Figure 10.34 shows some of the cracking through the I-beam at the stabilizer arm support. Since much of the cracking was centered at one of the stabilizer links or struts, it was decided to check the alignment of one link to another. Measurements indicated that one strut was posi-tioned $1\frac{1}{2}°$ off from the other struts. No cracking had been experienced on any of the struts positioned within $\pm\frac{1}{2}°$ of each other. Several other struts were off by $\frac{3}{4}°$ and the

FIGURE 10.34   Cracking of beam by stabilizer link.

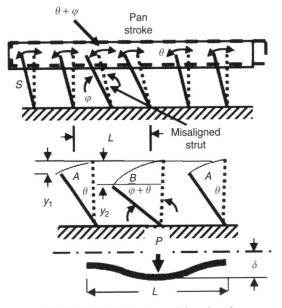

FIGURE 10.35   Drier model analyzed.

question was whether they had to be repositioned, which would require extended downtime. It was decided that a simple analytical model was needed to evaluate the effect of this misalignment.

Figure 10.35 shows the simple model analyzed, without all the links, counterbalance members, isolation springs, gussets, and drive points. The stabilizer links are represented by the dashed bars, and after they rotate they are in the solid bar position. When one of these links is off relative to the others, as the misaligned strut is, the beam is pulled down or pushed up, depending on the position. This is shown as δ and is the amount it is pulled down when the pan moves to the left. This exerts a bending moment on the structure and could result in a cyclic stress, as the structure vibrates at about 500 cycles/min. The question is: How does the misalignment angle affect the cyclic stress?

In Section 2.6 it was shown that the deflection δ and the moment M of a beam fixed at the ends and with a center load W or P can be determined by the equation

$$\delta = \frac{PL^3}{192EI}$$
$$M = \frac{PL}{8}$$

The moment in terms of the deflection is therefore

$$M = \frac{24EI\delta}{L^2}$$

By geometry the value of y_1 and y_2 can be determined at the end of the stroke knowing the angles and strut length. The pull-down or push-up deflection on the structure that will cause the bending moment M will be $\delta = y_2 - y_1$:

$$y_1 = S(1 - \cos\theta)$$
$$y_2 = S[\cos\varphi - \sin(90 - \varphi - \theta)]$$
$$\delta = y_2 - y_1$$

This δ value can then be inserted into the moment M equation. Data include $S = 12$ in., $E = 30 \times 10^6$ lb/in^2, $I = 63$ in^4, and $L = 48$ in. θ is the angle that the conveyor strut rocks through every cycle, and φ is the amount that strut B has been initially misaligned from the other struts, A.

From Table 10.6 it is clear that the less the misalignment angle φ is, the smaller the bending moment M will be. Since the stress is proportional to this bending moment, the cyclic stress will also be reduced. The stabilizer links have rubber bushings that will accept some misalignment, thus reducing the bending

TABLE 10.6 Effect of Misalignment on Bending Moment

Stroke Angle, θ (deg)	Misalignment Angle, φ (deg)	y_1 (in.)	y_2 (in.)	Forced Deflection, δ (in.)	Bending Moment (ft-lb)
5	1.5	0.046	0.073	0.027	45,000
5	0.5	0.046	0.055	0.009	15,000

moment. Since field data are available, the analysis need go no further. There has been no cracking on any of the driers when φ has been less than $\frac{1}{2}°$, but major failures have occurred when the strut misalignment was over $1\frac{1}{2}°$. There is good reason to believe that since there has been no cracking at less that $\frac{1}{2}°$, the stresses at this misalignment are $\pm 1,000\,\text{lb/in}^2$ or less, based on the discussion in Section 2.23 of the fatigue of welds. As things appear linear, the $1\frac{1}{2}°$ would be a stress of about

$$\frac{\pm 1,000(45,000)}{15,000} = \pm 3,000 \text{ lb/in}^2$$

Cracking could be expected at this level of cyclic stress.

All of the struts on the structure that were off more than $\frac{1}{2}°$ were repositioned and rewelded to within $\pm\frac{1}{4}°$, and the cracks ground out, rewelded, and stiffening plates added (just for comfort). Specifications for new conveyors stated that the stabilizer arms had to be within $\pm\frac{1}{4}°$, which the manufacturers could do with the correct fixturing.

Here's an example for which the exact stress didn't have to be determined since areas where misalignment-caused failures were known, as were areas where no failures occurred. After modifications, no further fatigue problems were experienced on these driers. As mentioned previously, vibrating conveyors and driers are excellent fatigue-testing machines. They are very sensitive to manufacturing tolerances, and detailed specifications are needed in the early design stage. All welded joints need to be of a fatigue-tolerant design and the cyclic stress kept very low, usually less than $\pm 2000\,\text{lb/in}^2$.

The simplicity of this analysis shouldn't undermine its importance. Many people were involved in this troubleshooting effort, as well as the equipment manufacturer. Other possible causes were mentioned, since the manufacturer didn't want to be liable for the extensive damage. Poor operating procedures, product overload, improper maintenance, wrong exciter speeds, too much product buildup, damage during installation, and storage procedures before installation, and stress corrosion cracking due to the atmosphere and product were all brought up during the troubleshooting.

One thing stood out, and that was a deviation. All the cracking centered about this deviation of a tolerance stack-up, causing the location of the strut to be off. The analysis only verified the consequence; it took trained specialists to notice the deviation. The information was not provided by the equipment manufacturer. The repair effort was shared with the manufacturer. The manufacturer provided the reengineering and parts along with an experienced field service representative to supervise the repair. The equipment owner supplied the plate materials, welders, and mechanics to implement the fixes.

No blame or warranty claims were directed at the manufacturer simply because litigation would put the manufacturer and plant personnel in adversary roles. This would have resulted in a further delay of the production schedule, along with difficulties in the procurement of new conveyors in the future. It is always better to work with the manufacturer to find a successful solution on an equipment problem, because once solved, it usually won't recur on new designs.

10.6.3 Case History: Oil Film Thickness of a Diesel Engine Bearing

Over the years the manufacture of a 4000-hp diesel engine was experiencing random failures on the highest-loaded 8-in.-diameter main bearing. The engine had been redesigned from a 2500-hp machine. Random failures had occurred at the lower horsepower level, too, but they were usually related to maintenance problems. It was decided that a check on the oil film thickness might be appropriate, since the main bearing load had increased by 60%. The crankshaft was robust enough so that fatigue failures were not a concern.

From Section 2.19, the following equation was provided for oil film thickness:

$$h_{min} = \frac{1}{2} c - \frac{3F}{\mu L (\text{rpm})} \left(\frac{c}{D}\right)^3$$

TABLE 10.7 Oil Film Equation Terms

h_{min}	Minimum Oil Film Thickness (in.)	Engine Data
μ	Oil viscosity (lb-sec/in²)	2.5×10^{-6}
c	Diametral clearance (in.)	0.006
F	Load (lb.)	50,000 and 31,250
rpm	Shaft speed	1100
L	Bearing length (in.)	8
D	Journal diameter (in.)	8

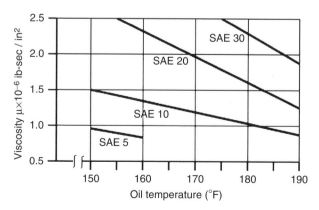

FIGURE 10.36 Oil viscosity–temperature curve.

The meaning and units are provided in Table 10.7. Figure 10.36, is useful when using the equation, as the units for viscosity are correct. The curves have been linearized. The results from the equation are:

Load, F (lb)	Minimum Film, h_{min} (in.)
31,250	0.00120
50,000	0.00012

The oil film thickness appears quite thin, so if the frequency of the random failures increases, they might be caused by the thinner oil film.

Use of this equation can be illustrated further by examining the effect of reducing the clearance from 0.006 in. on the diameter to 0.005 in. This increases the minimum film thickness h_{min} to 0.0008 in. from 0.00012 in. and was discussed further with a bearing manufacturer. A detailed analysis indicated that the smaller clearance could affect the oil flow around the bearing and the local oil temperature of the oil film. On reciprocating engines, oil films as thin as 0.0002 in. are acceptable since the load is only periodic per cycle and the film has time to reestablish itself. Nothing was done with the bearing design, and the random failures were acceptable; that is, they didn't cause warranty claims from customers.

On constant-directional loads such as on centrifugal compressors and turbines, the minimum oil film is normally between $h_{min} = 0.001$ and 0.003 in. Too much lower and the bearing may run at an elevated temperature or have metal-to-metal contact. Too much higher and stability problems may develop. Remember that this analysis was for 360° journal bearings and won't be applicable to tilting pad bearings or thrust bearings.

10.6.4   Case History: Leaking Flange Gasket

Flange and gasket analysis can be complicated, but sometimes a simple calculation can unveil the cause. Leaking gaskets are not unusual in industry and can be either an annoyance or have serious consequences. New plants usually have well-designed joints in critical applications, but leaks still happen. This case involves a class 150 6-in. flange with eight $\frac{3}{4}$-in. bolts which was found leaking after a system upset.

Gaskets usually leak for one or more of the following reasons:

1. The gasket clamping load or clamping stress is too low.
2. The gasket or joint is relaxed.
3. The gasket or flange surface is damaged or the flange is warped.
4. External forces are acting, such as piping thermal or alignment loads.
5. The wrong gasket material or design was chosen for the application.

The item that can be verified most quickly is the gasket load. When the pressure load opening the joint is larger or near the assembly clamping load on the bolts, a leak can be expected.

(bolt load − pressure load) must be greater than zero

TABLE 10.8 Leaking Flange Example Results

Condition	T (ft-lb)	p (lb/in^2)	σ_{gasket} (lb/in^2)	$\sigma_{\text{gasket}}/4p$
Design	40	300	1450	1.2
Upset	40	600	730	0.3
Increased torque	80	600	2910	1.2

It is useful to put this equation into terms of the nominal gasket stress, which means dividing the joint load by the gasket surface area:

$$\sigma_{\text{gasket}} = \frac{(60TN_{\text{bolts}}/d) - \pi D_{\text{inside}}^2 \times (p/4)}{\pi w \, D_{\text{inside}}} \quad \text{lb/in}^2$$

From the equation it is obvious that the joint sealing stress increases with the bolt torque T and the number of bolts N_{bolts} clamping the flange together. It decreases as the gasket width w increases or the internal pressure p increases. With good flanges and gasket material and no external loads, a reasonable value for flat gasket stress where no leak should be expected is about four or more times the internal pressure. Rubber gaskets are closer to one times the pressure. Remember that these are not rules to design by, only methods to troubleshoot by.

For this case $w = 0.625$ in., $d = 0.75$ in., $D_{\text{inside}} = 6$ in., and Table 10.8 can be developed. It can be seen from the table that at design conditions, the joint has sealed. At upset conditions of double the pressure, the joint leaked. There were many problems. The flange was of the wrong type, the gasket was too wide, and the bolts were not tightened adequately. Increasing the bolt torque alleviated the leaking.

The stress in the bolt needed to be checked to ensure that it didn't exceed the yield strength of 105,000 lb/in^2. The following equation is for unlubricated threads:

$$\sigma_{\text{bolt}} = \frac{120T}{d^3}$$
$$= 22,760 \text{ lb/in}^2$$

Fifty percent of yield stress is usually acceptable, but in a thorough analysis the flange robustness should also be verified. At elevated temperatures, the yield stress of the bolts and flange at temperatures will have to be used. The process temperature was not a factor for this case. Also, the gasket stress cannot be so high as to crush the gasket.

Hot-bolt torquing, tightening the flange bolts further to stop the leak when the unit is in operation, can be dangerous. The leak may have corroded or cracked the stud bolts, which could break off and result in a much larger leak. Every company should have a hot-bolting procedure to follow that checks the condition of the studs and addresses the risks before proceeding. When hot-bolt torquing doesn't work and the flange cannot be isolated or taken out of service, consultants who specialize in injection sealing, clamping, banding, or boxing of the joint should be utilized. Any of these temporary sealing procedures may increase the thermal stress on a joint and therefore require a sound engineering approach.

11

FITNESS FOR SERVICE, WITH CASE HISTORIES

With the wide scope of activities that engineers and technicians are expected to perform, few have the luxury of becoming specialists in a single area. We are usually expected to be able to address any equipment problem. This can be dangerous when the subject is as critical as making decisions as to whether a piece of equipment can continue to be operated. Fitness for service is an area where you need to know what it is about but you don't have to know how to do it. There are experts in this area who should be used for these types of calculations. This chapter can be used as a brief overview of some of the types of activities and problems that can be addressed. It should not be used as a design guide.

In the process industry, pressure vessels, piping, and tankage that have been in service for 10 or 20 years usually deteriorate in some manner. Plant engineers and supervisors don't usually enhance their careers by telling the plant manager continually that equipment has to be shut down to make repairs or, worse yet, replaced. It is much more productive to be able to analyze the piece of equipment and develop an approach to operating it safely to a planned downtime, rather than contributing to a forced outage.

Pressure vessels in hydrocarbon service are built and repaired to various country codes, standards, and guides. Many companies have fitness for service (FFS) guides for extending the life of equipment that has suffered some type of degradation mechanism. The American Petroleum Industry (API) has published API 579 [15] for this purpose. It considers corrosion, pitting, blisters, and cracklike defects. Safe operation requires that detailed FFS be performed and documented using accepted techniques

Analytical Troubleshooting of Process Machinery and Pressure Vessels: Including Real-World Case Studies, by Anthony Sofronas
Copyright © 2006 John Wiley & Sons, Inc.

on a piece of equipment for continued operation. However, simple case history studies still have their place as a guide.

In this section case histories are based on two modes of deterioration that limit the service life: local corrosion and cracklike defects in a vessel. In each case the question is what needs to be done so that the vessel can remain in operation. In each case, the answer that no one wants to hear is that the vessel needs to be replaced.

The case histories here are in an abbreviated form; the actual solutions were much more exhaustive. Much of the energy was involved in obtaining the thinned dimensions of the thinned area and characterizing the minimum thickness and the cracklike defect. The examples here have been greatly simplified and are for illustration purposes only. You are again cautioned to follow industry standard methods when performing such analyses, and to seek the advice of experts in these areas on your specific problems if your experience is limited. Two methods are discussed for analyzing these two modes of failure.

For locally corroded areas, the principal concern will be if the corroded region has exceeded the design thickness to contain the design pressure. If it has, the method used will be to examine the *remaining strength factor* (RSF). This is the strength required to limit plastic collapse of the vessel. A value of 0.9 RSF often means that the plastic strain allowed in the thinned region has been limited to 2%. An RSF value of 1 would mean that the strength has not been reduced from the design condition. Lower RSF values would indicate higher risks and would have to be evaluated on the basis of the risk that could be tolerated.

For the cracklike defect example, a failure analysis diagram (FAD) will be used. This analyzes the vessel on the basis of the tendency of plastic collapse of the vessel and the tendency toward a brittle fracture of the vessel. It is concerned with temperatures, material toughness properties, stresses, and defect size.

In both of these methods there are many limitations and conditions that must be met before they can be used. Addressing these conditions is usually the most important part of the analysis and is something not in the scope of these examples. The various standards guide you through these conditions before allowing you to proceed with the analysis. For example, several cracks may have to be characterized as a single crack, or cyclic loads or a crack near a discontinuity may not be allowed. In the case of uniform corrosion, the question is what constitutes this condition; that is, how do you get a surface with pits and variations into a thinned thickness? This is not a simple question to answer and is one of the reasons for the many checks that are necessary in the analysis methods. Become familiar with the standard or method used in your region or country.

11.1 A LITTLE ABOUT CORROSION

Why include a section on corrosion? Corrosion is a complex subject requiring years of experience in metallurgy and an understanding of materials. It is a discipline all by itself and one on which this author is not an authority. Actually, I usually rely on experts in this area. The reason we discuss corrosion is because it is so important in model building. It is how we have verified the mode of failure. When we build a mechanical engineering model where we have had a failure due to stress and the model fails to predict the cause, it is usually because some other mechanism is at work. In the majority of failures analyzed in the processing industry, that mechanism is corrosion. So the intent of this chapter is not to teach you all about corrosion but to illustrate why it is important for you to consider it.

A colleague of mine was towing his rather expensive boat out of the water and onto a trailer. The trailer failed because the tubular frame buckled, as it had rusted from within because a drain hole had plugged. His boat was badly damaged. Old trailer, new boat. The trailer was designed correctly, but the original tube thickness of $\frac{1}{8}$ in. had rusted to $\frac{1}{64}$ in. The tube buckled and the boat dropped. The same thing can happen to piping and vessels in the petrochemical industry, except that corrosion can be from the inside or the outside. To complicate things, there are many corrosive products, some of which can corrode through the wrong materials in a matter of days, with significant consequences.

11.2 STRESS CORROSION CRACKING

Corrosion can make traditional strength, fatigue, and brittle fracture analysis methods totally unreliable when stress corrosion cracking (SCC) is present. This is because cracks have been generated by the corrosion mechanism and make analysis methods very unpredictable unless extensive test data are available. Since cracks are already present, the traditional material properties are highly questionable. Strengths and toughness (i.e., brittle fracture resistance) are much lower when SCC is present.

For example, a fatigue failure had occurred on shafting from a high-horsepower gearbox. Traditional fatigue methods predicted that the loads, considering surface finish and other factors, would not have caused such a failure. Even with extreme load conditions, no failure was predicted. But the shaft did fail in fatigue. A metallurgical examination determined that fretting corrosion due to a sleeve rubbing on the shaft had reduced the surface fatigue strength so that small cracks formed. The cracks then grew in fatigue to result in a final failure. Fretting corrosion is discussed in Section 9.14. There are all sorts of corrosion, and most materials can be affected by some form of corrosion.

When you work in the petrochemical industry and you are involved with troubleshooting, not a day goes by that you don't hear about stress corrosion cracking. When sending a part to the materials laboratory, one can always expect to hear that SCC was prevalent or the cause of the failure. The author once suspected that outside materials laboratories had a generic letter that left blank the part being investigated. It then goes on to describe that it was caused by SCC.

When you review the literature, it becomes pretty obvious that no matter how exotic a metal is, in certain environments SCC can occur. Even what may seem to us as rather benign environments, such as seawater, can cause SCC in aluminum, carbon, alloy, and stainless steels, because of the chloride content. Inconels, monels, and nickels can develop SCC with certain caustics. Even gold, lead, and titanium can suffer SCC under certain conditions.

Eliminating SCC is quite simple. Either eliminate the stress or eliminate the environment. The problem is that neither of these is easy to eliminate. Stress may be tensile residual stress, and the environment is usually something you can't change. It could be in the process itself. Methods are available to prevent it: for example, putting compressive stresses on the surfaces by shot peening, stress relieving to reduce tensile stresses, or coating the surface to shield it from the environment. These solutions are best left to experts in the area, since if done incorrectly, cracking and failures can be expected.

11.3   UNIFORM CORROSION

Other types of corrosion, such as uniform corrosion, are also quite common and are easier to deal with. Rusting of carbon steel piping and vessels can occur due to corrosion under the insulation (CUI). Moisture forms or rain collects and oxidizes the metal. Some metals, such as aluminum, form their own protective films on the surface, as you can see if you scratch an aluminum surface. An oxide forms and the bright scratch turns gray.

Fortunately, uniform corrosion of carbon steel is fairly predictable and rather easy to control. Inspection intervals can be set and vessel and piping life can be determined based on the corrosion rate in a stable environment. Cleaning by blasting or grinding, painting, and coating along with a fitness for service evaluation can extend service far beyond the vessel's original design life. Selection of a material such as stainless steel can eliminate the oxidation problem. When all else fails, and with care and careful planning, repair sections can be welded in or corroded sections can be built up with weld material.

11.3.1   Case History: Local Corrosion of a Vessel Wall

The following case history occurred on a pressure vessel that had been in service for 20 years. Insulation was removed and spot inspection was performed while the vessel was in operation. Inspection records indicated that uniform local corrosion had occurred at a rate of 0.0075 in. per year, which exceeded the allowable vessel design thickness. A downtime was planned to occur in a year, and the question asked by management was if continued safe operation to the downtime was possible. Figure 11.1 shows the vessel and the locally thinned area (LTA).

The following analysis method is not unlike many of the fitness-for-service techniques now being utilized. It is used here because of its simplicity. The basis of the method is developed in a technical paper [52] that is readily available. This is an ideal case as there were no nozzles or other defects, such as cracks or pits, in the corroded zone. Also, the material was ductile and the thinned thickness was uniform. There were no loads other than the pressure load to consider. However, all of these types of complications and more could have been analyzed using other fitness-for-service methods. Such detailed analysis is technically challenging and should be performed

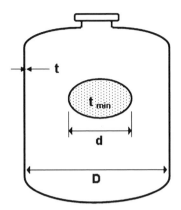

FIGURE 11.1 Locally thinned vessel.

by engineers well qualified in the use of fracture mechanics and in performing and understanding stress analysis methods.

In the following equations, t_{min} represents the thickness in the LTA and t is the thickness of the unthinned area. RSF is a measure of the plastic collapse resistance of the vessel. For our purposes and for many fitness-for-service approaches, the RSF should be equal to or greater than 0.9. The equations follow.

$$z = \frac{t_{min}}{t}$$

$$A = 55(z)^4 - 168(z)^3 + 189(z)^2 - 100(z) + 25$$

$$m = 1 + 2.3z \times \left(\frac{Ad}{D}\right)^{2.3}$$

$$RSF = \frac{z}{1 - (1/m)(1 - z)}$$

For the vessel being considered, where $d = 24$ in., $D = 120$ in., $t = 0.75$, and $t_{min} = 0.6$ in.,

$$RSF = 0.937 > 0.9 \qquad OK$$

Even with this idealized solution, some valuable information is now available. The simple analysis indicates that the vessel can continue to operate safely until the scheduled downtime. An added benefit is that it appears that the vessel does not have to be repaired at that time, as the thinned area only needs to be cleaned and protected. By using better insulating procedures, the future corrosion rate can be greatly reduced or eliminated. Inspections can be scheduled based on these new expected corrosion rates and local standards. Reporting to management that the equipment can be operated without interruption until the downtime, and that the expense will be minimal is a pleasant experience. Unfortunately, all problems do not turn out this well.

11.4   PITTING CORROSION

The pitting considered here is not that caused by subsurface stresses such as occurs in bearings, gearing, or other highly loaded surfaces. Rather, this is corrosion related and is often found on pressure vessels and piping. Here, a *pit* is defined as having a diameter on the order of the vessel thickness and a depth that is less than the thickness. The danger is that a pinhole leak may develop from a pit if its depth is close to the wall thickness. With many such pits in a localized area, the containment wall can be weakened. There are many causes for pitting-type damage, such as chloride pitting of stainless steels or CUI of carbon steels. Buried piping usually suffers from some sort of pitting corrosion when the surface treatment that acts as a moisture barrier fails or the cathode protection system becomes ineffective. Fluid-related cavitation damage sometime results in a surface that has a pitted appearance. Causes and remedies for corrosion-related problems are best left to a corrosion engineer, who is usually part of a materials laboratory's team. Fitness-for-service analysis and repair methods are available from one of the several industry standards.

11.4.1   Case History: Pitting Corrosion of a Vessel Wall

Like the uniform corrosion case, this corrosion was caused by CUI. The pitting is distributed in a circle with a diameter of 50 in. There are 100 pits with diameters of approximately $\frac{1}{4}$ in. They are spaced about 1 in. apart and range from 0.25 in. deep to a maximum of 0.6 in. deep. The pitting is only on the outside of the 1-in.-thick vessel.

As in all things, in engineering it is usually best to try the easiest approach first. If it doesn't work, one can proceed to the more difficult. So it is with pitting. API 510 includes some screening procedures to check to see if a specific amount of pitting is acceptable. Widely scattered pits can be screened by checking that the depth of the pit/wall thickness is not greater than 0.5, exclusive of the corrosion allowance. The total pit area should not exceed 7 in² in an 8-in.-diameter circle. The sum of these pit diameters in the circle should not be greater than 2 in. Obviously, this is not true if the cause of the pitting corrosion has not been stopped, and the pits can continue to grow. When the pits are close together, the area will start to be a locally corroded area.

- 0.6-in. pit/1.0-in. wall > 0.5—fails
- The pit area in the 8-in.-diameter circle is < 7 in²—pass
- The sum of the pit diameters in the circle is > 2 in.—fails

So for this case the pitting is unacceptable according to API 510.

A second check would be to see if the design thickness of the vessel has been exceeded. This is done by calculations using the code to which the vessel was designed. In this case it was to ASME Div. 1.

$$t = \frac{pR}{s}$$

For this vessel, $p = 200$ psi, $D = 50$ in., $E = 1$, and $s = 17,500$ psi, so

$$t = \frac{200(25)}{17,500} = 0.28 \text{ in.}$$

It seems that there was a lot of extra material in this vessel's wall, and a thickness of only 0.28 in. is required to contain the pressure. Since the thinnest part of the wall is $1.0 - 0.6 = 0.4$ in., the deepest pit is probably okay.

The difficulty is determining the minimum thickness of this zone. This is true especially if corrosion is from the inside and outside of the vessel. How to analyze this becomes even more complex when not only is pressure stress present, but also bending stresses. It becomes a problem similar to that of analyzing a partially perforated plate, which is not an easy one to solve. Advanced analysis techniques are required under these conditions, such as finite-element analysis. API 579 provides a more detailed approach to follow when safety is an issue.

11.5   BRITTLE FRACTURE CONCERNS

Although not a frequent occurrence, vessels do fracture, and when they do, catastrophic failures can occur, usually in welds. They occur without plastic deformation and at extremely high speeds, on the order of 7000 ft/sec. They can occur at stress levels below those of general yielding. Although this subject is quite complex and much too critical to apply rules of thumb to, it is important for those who troubleshoot failures to be aware of this potential concern.

The basic principle is that a crack or defect is assumed to exist in the material or weld. The material's toughness has to be able to withstand the stress intensity at the tip of the crack under all stress and temperature conditions. Temperature is critical since the toughness of steels changes drastically at low temperatures. Vessel designers know this and the construction codes take it into account when specifying materials for an application.

Figure 11.2 represents the temperature of an A36 steel and the resulting toughness as measured by a Charpy V (CVN) impact test. For the purpose of discussion, steels with a CVN value of 15 ft-lb or greater at temperature will be considered unlikely to suffer a brittle fracture failure when good design procedures have been followed. Notice on the graph that at 100°F, this steel appears to have adequate impact absorption capabilities at 30 ft-lb. Drop the temperature to 0°F and the material is brittle. Notice the transition region, where this steel's ability to absorb energy drops quickly.

K_{IC} is a material property that is a measure of the material's toughness. For steels, when no data are available, many use the correlation

$$K_{IC} = (150 \text{ CVN})^{1/2} \text{ ksi (in.)}^{1/2}$$

Large K_{IC} values mean tougher materials. K_{IC} represents a material's ability to withstand a given stress field at the tip of a crack (i.e., its ability to keep the crack from propagating). This is test data information and has the strange units ksi (in.)$^{1/2}$.

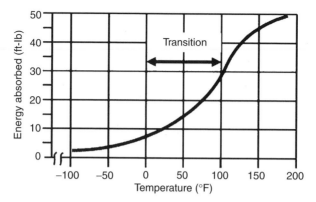

FIGURE 11.2   Toughness curve for an A36 steel.

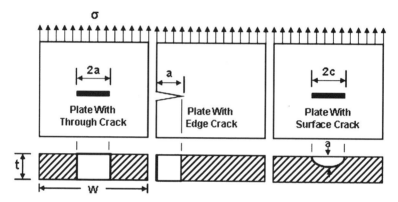

FIGURE 11.3   Various crack geometries.

The second term of importance is K_I, the stress intensity factor, a calculated value. It represents the stress field intensity at the tip of the crack. It depends on the crack size, crack geometry, and stress field trying to open the crack. There are many equations for different crack profiles, but for this discussion a simple crack through a plate will be considered. The deceptively simple K_I equation for these geometries is

$$K_I = \sigma F \, (\pi a)^{1/2} \text{ ksi (in.)}^{1/2}$$

Knowing the nominal stress σ (ksi) trying to open a crack of length a, $2a$ or $2c$, and the shape factor F, K_I can be calculated. This is valid only when the width of the plate is large compared to a, $2a$, or $2c$.

The surface crack on the right in Figure 11.3 is a "thumbnail" crack [53] of width $2c$ and depth a. It is assumed for our cases to grow in the ratio

$$\frac{2c}{a} = 6$$

This means that if a is calculated to be 0.125 in. deep, the length of the crack ($2c$) would be

$$2c = 6a$$

Deceptively simple means that if these equations are used without knowledge, the results can be significantly in error by orders of magnitude. That's why this section is for educational purposes only and not to be used to make critical brittle fracture or growth rate decisions.

As long as the following condition is met, brittle fracture should not occur and the crack should be stable:

$$K_{IC} > K_I$$

This simply says that the material's toughness to resist the stress intensity needs to be greater than the stress intensity developed at the tip of the through crack.

11.5.1 Academic Example: Temperature Effect on Steel Plate

A plate made of A36 steel has a through crack length of $2a = 1.0$ in. It usually is in 100°F service but is going to be used in a 0°F environment. Can you think of any problems?

From Figure 11.2, the following data are obtained:

Temperature (°F)	Charpy V (ft-lb)	Toughness, K_{IC} [ksi (in.)$^{1/2}$]
100	30	$(150 \times 30)^{1/2} = 67$
0	8	$(150 \times 8)^{1/2} = 35$

The nominal stress is known to be 30 ksi, which consists of 15 ksi of applied stress and 15 ksi of residual stress due to lack of stress relieving. For this case the shape factor $F = 1.0$. At 100°F:

$$K_I = \sigma F (\pi a)^{1/2} = 30(1.0)[\pi(0.5)]^{1/2}$$

$$= 37.6 \text{ ksi (in.)}^{1/2}$$

$$K_{IC} > K_I$$

$$67 > 37.6 \quad \text{Yes!}$$

Since the inequality is met, there should be adequate toughness.

At 0°F:

$$K_{IC} > K_I$$

$$35 > 37.6 \quad \text{No!}$$

The inequality is not met, so there is not adequate toughness and a brittle fracture could occur if a crack were to go through the vessel. What might be done to correct this problem? Obviously, fixing the crack in an approved manner would help, but what would happen if the plate could be stress-relieved? The stress would drop to 15 ksi and K_I would be around 19 ksi (in.)$^{1/2}$, which is less than 35 ksi (in.)$^{1/2}$ and may be adequate. Not many engineers would use this material for low-temperature service in critical applications, and national codes wouldn't allow it.

Why would one be interested in a through crack in a pressure vessel? The reason is that if a crack were to develop through, the designer would want it to be a stable crack. Hydrocarbon detectors on the unit or unit personnel would notice a small leak and shut the unit down in a controlled fashion. In an unstable leak, such as that which might occur in the 0°F case, the through crack could crack open the vessel, allowing an uncontrolled leak in less than 1 second. The first is a much more feasible failure mode and illustrates why pressure vessel steels and designs are so well controlled by codes.

Not much more will be said on this subject except that problems should be left to experts in this area. The simple analysis shown can cause many problems if not used correctly and can produce wrong answers. Crack geometry, toughness variations in welds, residual stresses, stress distribution, corrosion effects, and more must all be understood by the expert performing the analysis. The example does show that there are methods available to determine if brittle fracture might be a concern and will allow the technician to understand what the expert is talking about.

Partially through cracks or through cracks can be analyzed using fracture mechanics principles. How fast a nonbrittle crack will grow under cyclic loading can also be evaluated, which is useful for life estimates. Below a certain threshold, stress cracks arrest and will not propagate. This makes sense, since intuitively one would feel that there should be a $\Delta\sigma$ value small enough that the crack will not grow.

11.5.2   Case History: Crack Like Defect in a Vessel Wall

The same vessel as used in Section 11.3.1 is considered here. In this case a crack has also been located along a vertical weld after the insulation was removed. This crack is in a noncorroded portion of a 0.5-in.-thick vessel and is 1.0 in. long and 0.2 in. deep in the vessel wall, parallel to a weld. Can the vessel continue to operate for one year until its next scheduled downtime?

Some questions are in order before we begin our analysis. First: Is the crack growing? Any growth in a crack is critical, and the risk must be analyzed immediately or the unit shut down if the cause is unknown. This is suggested for several reasons. First, if there is a corrosion mechanism such as stress corrosion cracking in progress, it will be extremely difficult to determine the toughness of the material and thus the potential for a brittle fracture. Second, if a brittle fracture can occur, it can "unzip" a vessel (i.e., run along a weld) at more than 7000 ft/sec, and this doesn't allow any time to monitor the growth.

Another question might be: "What is the possibility of other external or internal cracks this size or larger?" Without recent inspection records, only further inspection will answer this. Again we are into risk analysis. Obviously, the risk that one is willing to take with a water tank in the middle of a prairie is quite a bit different from that for a vessel inside a plant with hydrocarbon at 300 lb/in² pressure. Let us say that this is a simple problem, and that because of product type and location, risk is not a concern, but it would be detrimental to operations to have to shut down the unit. Also, this is the only crack detected. In such a case, we can continue with our simplified analysis.

A little more needs to be known about this vessel. The material it is constructed of is an old 1940s' version of A212 steel. From information tabulated [53] on this steel, its toughness near a weld is Charpy 5 ft-lb. This is not a very tough steel, so it would have to be treated as a brittle material. The stress component trying to open this crack is the hoop stress of the vessel, which results in a primary membrane stress of 16,000 lb/in². Postweld heat treatment had been performed on the vessel and it has been hydrotested.

Failure assessment diagrams (FADs) are used to assess the potential for a structure to fail from a brittle type of failure or a plastic collapse such as is experienced when a structure exceeds its yield strength. A material may be tough enough not to have a brittle fracture, but it may fail by yielding, and the FAD will indicate this. In the FAD, K_r is a measure of the toughness and S_r is a measure of the stress applied to the plastic collapse stress. When a material is very tough and the stress is high, the structure will usually fail by yielding when $S_r = 1$. A brittle material fails when $K_r = 1$. The details on calculating the K_r and S_r ratios are available from other sources [e.g., 54] and are not reviewed here.

For this case history the FAD coordinates shown in Figure 11.4 were $K_r = 0.71$ and $S_r = 0.44$, which fall well inside the diagram, in the safe region, so this particular crack is acceptable. Were the crack unacceptable, the crack would have to be ground out and the vessel reanalyzed for this thinner section. In the extreme case, the crack may have to be welded using the special precautions required when grinding and welding brittle materials. This would be a good time to contact your welding and material experts, as such repairs can sometimes be of more concern than the crack was. Had this vessel not been made of PWHT to avoid residual stresses, it probably

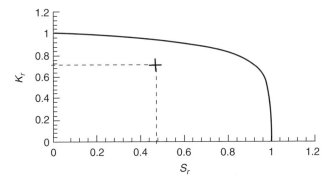

FIGURE 11.4 Failure assessment diagram.

would have fallen outside the curve in the unsafe region on the FAD (Figure 11.4). This is the problem with many old vessels. They have been there for 40 years, have cracks in them, but have not failed in brittle fracture. The question usually is why they are considered unsafe. Most of the time this is because we just don't know enough about the vessel. The analysis procedures have many built-in safeguards. The stress distribution, material toughness, or crack geometry may be such that the crack just won't propagate or is arrested in nonhomogeneous tougher material. The analysis method may not be accurate enough for such a complex mechanism. Whatever the reason, care must be taken so that vessels are not condemned because of the analysis alone. Samples may need to be tested for the actual toughness, and risk-based methods may need to be employed and experienced advice sought before the appropriate engineering decision can be made. The analysis helps in making the decision, but alone is usually not enough.

11.6   COLD SERVICE EVALUATIONS

Pressure vessels can be rather unexciting pieces of equipment compared to more complex pieces of machinery. Machines seem to have their own personalities, whereas pressure vessels usually lead long, uneventful lives. Specialists who have experience with both realize that when problems do arise with pressure vessels, they are anything but dull. High-pressure, volatile, and/or hazardous products, coupled with safety, environmental, and production losses, are the reasons why pressure vessels are designed and inspected to various state and country codes. Material properties, fabrication techniques, and quality control are carefully controlled.

One area that can result in catastrophic failures in pressure vessels is the brittle fracture of materials. The materials of construction must have suitable toughness for low temperatures. Low-toughness materials with a defect such as a crack and high enough stress can have a brittle fracture failure. Glass is a brittle material. Glass can hold water just fine but it shatters if over stressed. This can happen to brittle metals as well when overstressed at a defect.

This case history is not one of a brittle fracture failure. Fortunately, such failures are rare. The present code toughness standards are quite good at specifying materials with adequate toughness at the design temperature. When the code and UCS diagrams are mentioned, the ASME pressure vessel code is being followed [55]. Charpy impact testing is a measure of toughness in ft-lb. A material value of 20 ft-lb or higher, at temperature, usually indicates a tough material. Older steels can have Charpy numbers of 3 ft-lb or less, which would be considered quite brittle. Although the probability of a brittle fracture failure may be low, the consequence of such a failure is usually so high that every effort is made to reduce the probability as much as possible.

11.6.1   Case History: Cold Service Vessel

In this case history, a 40-year-old pressure vessel was being reviewed for its suitability for a new service. The drum contained liquid ethane that could refrigerate (i.e., as the vessel is depressurized, the vessel temperature would decrease). The abnormal operating scenario being evaluated was that of overpressurization. It was assumed that under this condition the safety valve would blow, reducing the pressure to a safe level. However, it is also quite possible that the safety valve may not reseat itself immediately. Under this scenario the vessel could get quite cold, due to the depressurization. If it then reseats and is brought quickly back up to the design pressure, an unsafe condition could occur. This is what is being reviewed.

Minimum design metal temperature (MDMT) is a code term. Above this temperature the material is considered tough enough by code standards at the design pressure. Below this temperature, the pressure must be reduced to meet the toughness requirements of the code. Additional temperature credits are available for vessels with postweld heat treatment or when a vessel thickness is more than is required to meet the code allowable stress. In this case history, only pressure reduction is considered.

The MDMT of the material is selected from UCS-66 for the type of material and thickness of that material. The governing thickness of the example in this case study is 0.4 in., so the MDMT is about 18°F for curve A material, which is the only curve shown in Figure 11.5. This is generally the curve used for older materials. The governing thickness for welded plates is usually the thinner of the two joined. For non-welded plates it is usually 25% of the thickness of the plate. Above this temperature impact testing of the material is not required as long as the design membrane stresses are not exceeded, which means that the material is tough enough to meet the code requirements. However, when the design stresses are lower than the allowable design stresses, the MDMT are allowed to be lower. This is good since during depressurization the metal temperatures get colder, but the pressures are also dropping. Basically, the material may be more brittle, but the stress, which could drive a brittle fracture, is also decreasing. At a low enough stress it has been shown that there is not enough stress to drive a brittle fracture failure. For example, there can be deep notches in a glass holding water, but if there is no stress on the glass, it will not break. Heat up or cool down the glass suddenly, and it may break, due to the increase in the thermal stresses at the notch. The material was not ductile enough to support the strain, so failure proceeds in a brittle manner.

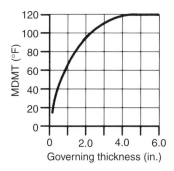

FIGURE 11.5   Minimum design metal temperature.

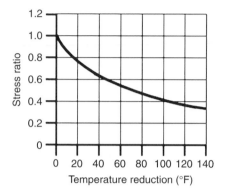

FIGURE 11.6   Stress ratio temperature reduction.

UCS 66.1 is the curve that considers the lower stresses and is used to determine further reduction in the MDMT due to the reduced stress. It is called the *stress ratio temperature reduction curve* and is shown in Figure 11.6. The stress ratio is usually calculated as the operating pressure divided by the design pressure. At the design pressure (psig) the ratio is 1. Notice in this example that only the shell wall thickness under membrane stress is being considered. In an analysis such as this, all critical components must be considered. For example, the flanges of this vessel are undergoing a bolt load in addition to the pressure load and will have to be considered separately, as will the bolt. With horizontal tanks supported on saddles, which are sometimes called *bullets*, the bending loads will also have to be considered.

Since vessel wall thickness and membrane stress are directly related to the maximum available working pressure (MAWP), they can be scaled down by the ratio and the temperature reduction determined. The temperature reduction is then subtracted from the MDMT as shown in Table 11.1. The resulting value can then be plotted to show the allowable pressure–temperature curve. This curve, sometimes called the *minimum safe operating temperature curve* (MSOT), is shown in Figure 11.7. Points that fall under the dashed MSOT curve are expected to meet code toughness requirements for the shell of the vessel under membrane stress only.

TABLE 11.1   Temperature Reduction Due to Stress Ratio

Stress Ratio, UCS 66-1	ΔT (°F)	MDMT $-\Delta T$	Pressure (psig)
1.0	0	18	80
0.9	10	8	72
0.8	20	−2	64
0.7	30	−12	56
0.6	40	−22	48
0.5	60	−42	40
0.4	105	−87	32

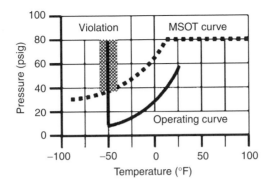

FIGURE 11.7   MSOT and depressurization.

The depressurization curve is also shown, as is the repressurization. Notice that there is a violation of the curve as it exceeds the MSOT curve (shown as the shaded area). There are many ways that this could be corrected, such as through additional operator training or by modifying the repressurization procedure either by automating the controls or in the operating procedures. In this case history, the solution was a set of design procedures that contained many of the parameters needed to effect a change and allow the system to continue to operate.

This analysis has been for discussion purposes and is not complete. These types of problems can become complicated with nozzles or attachments, which may result in bending stresses being introduced into the vessel shell. Flanges, heads, supports, and bolts also require consideration, as do heat-up and cool-down rates and the thermal stress they can introduce. Re-rating vessels such as this is not a trivial exercise and is best left to engineers who specialize in this area.

11.7   CRACK GROWTH AND FATIGUE LIFE

The work in Section 2.5 on the fatigue of structures, especially the modified Goodman diagrams, was based on fatigue endurance limits. These were based on polished specimens, and most of the lifetime was spent developing the initial crack. They are not of much use when a crack already exists. There are methods that take the crack size into consideration.

Although this is a fatigue-related subject, it is included here since it utilizes nomenclature from the fracture mechanics section. In this section the interest is in how many cycles or how long will it take for an existing crack to grow from its initial length to some final length. The final length could be a critical length for a brittle fracture or a length that reduces the lifetime by reducing the area. For example, it may reveal the length of time it will take a through crack to grow to 10 times its original length. Many computer programs are available to do these types of calculations, on all geometries, materials, stress fields, and crack sizes, but for example purposes, a simple closed-form solution will be used.

In fracture mechanics, the crack growth rate, how much the crack will grow per load application, can be represented by the *Paris differential equation:*

$$\frac{da}{dN} = C(\Delta K)^m$$

where da is the change in crack length (in.), dN the change in cycle, and C and m are material parameters. The stress intensity range is $\Delta K_1 = K_{max} - K_{min}$ and characterizes the cyclic stresses and strains ahead of the crack tip. From test data on stainless steels, this relationship is approximately

$$\frac{da}{dN} = 3.0 \times 10^{-10} \, (\Delta K_1)^{3.25}$$

For many carbon steels,

$$\frac{da}{dN} = 3.6 \times 10^{-10} \, (\Delta K_1)^{3.0}$$

$\Delta K_1 = \Delta\sigma(\pi a)^{1/2}$ is for a plate with a through crack, as shown in the brittle fracture section.

ΔK_1 will be substituted into the crack growth equation for stainless steels, since an actual example with this type of steel is used in Section 10.6.1:

$$\frac{da}{dN} = 3.0 \times 10^{-10} \, [\Delta\sigma\,(\pi a)^{1/2}]^{3.25}$$

$$= [19.27 \times 10^{-10} \, (\Delta\sigma^{3.25})] \int a^{3.25/2}$$

This can be separated and integrated to determine the number of cycles for the crack to grow from its initial length a_0 to a final length a_f:

$$\int dN = [19.27 \times 10^{-10} \, (\Delta\sigma^{3.25})]^{-1} \int \frac{da}{a^{3.25/2}}$$

$$N = \frac{8.3 \times 10^8}{\Delta\sigma^{3.25}} \left(\frac{1}{a_0^{0.625}} - \frac{1}{a_f^{0.625}} \right) \qquad \text{cycles}$$

With the initial and final dimensions known, the number of cycles to grow the through crack from a_0 to a_f can be determined.

For example, consider the time for a crack to grow in a piece of stainless steel with a through plate crack geometry as in the previous example. Data for this case are $2a = a_0 = 0.5$ in., $a_f = 2.0$ in., $\Delta\sigma = 6$ ksi, and

$$N = 2.19 \times 10^6 \text{ cycles}$$

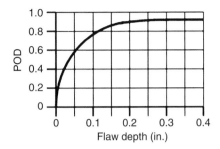

FIGURE 11.8 Probability of detection, UT, and MPT.

11.8 FINDING THOSE CRACKS

In this chapter we have talked about the importance of locating cracks and characterizing their size and shape. We do not describe how to perform these inspections. Many methods, such as visual, dye penetrants, radiographic testing (RT), ultrasonic testing (UT), and magnetic particle inspection (MPT), are used to define the size of cracks. All require considerable training, experience, and skill to use and interpret. When performing a fitness-for-service evaluation, it is all too often the specialist who says: "This is a brittle vessel and a 100% inspection is required to ensure that there are no cracks." But a 200-ft-diameter tank 40 ft high is a formidable surface area to inspect for 1- or 2-in.-long cracks on the inside and outside. That represents more than 1 acre of metal the inspector has to survey. Also, the probability of detection (POD) graph of Figure 11.8 shows that for UT and MPT, there is a low probability that all crack sizes can be found based on the skill of the operator and the accuracy of the equipment to detect the crack. Curves such as this make safety experts nervous, since they suggest that not all critical cracks can be located by traditional means.

Methods such as acoustic emissions, which "listen" to the sound of a crack propagating under load, are sometime used to determine if growing cracks are present in a vessel. Obviously, the vessel has to be pressured higher than it has been in the past to have a crack grow, which is also a concern. So this testing procedure needs to be used with extreme caution on brittle materials. Most brittle fracture failures have occurred during hydrotesting, which is normally the time that such a procedure would be used.

The technical specialists involved in the science of inspection methods and procedures are different from the specialists who perform analysis and repairs. Rarely does one person have the knowledge to perform all of these functions with a high degree of proficiency. It is wise for an organization to develop specialists or utilize consultants who have experience in all these critical areas.

11.9 TROUBLESHOOTING ISN'T EASY

As a closing to this book, this is a precaution that was printed in a series [56] I have written on troubleshooting case histories. Much time was spent in the series, as in this book, on analytical troubleshooting: using mathematical models to help determine the cause of failures. How to prevent them from recurring was the ultimate goal.

The author has often received comments to the effect that these case histories can't be too difficult when it took only one page in a magazine article to solve each of them. Although I could reply simply that I was pleased that I had made things so clear, I'm afraid there may have been a misconception on the effort that was required. Most of the case histories in this book took more than a month to solve and to convince management of the cause. It sometimes took over a year to obtain the funding to implement a permanent solution.

The troubleshooting sessions relied on many in-house experts, such as engineers, machinists, operators, and inspectors and were true team efforts. The high-visibility failures, those involving safety or lost production, required rather intense daily status meetings with management. In almost all cases there was not just one cause, but other less obvious causes that also had to be addressed. For example, consider a shaft that fails because of a torsional overload. Calculations show that this was an overload at startup. Other causes might have been that there were no startup procedures, that the procedures were in error, and/or that the procedures were not complete, or possibly, were not followed. A problem-solving session has to consider and address all the causes.

As discussed at length in Section 8.3, problem solving comprises the following stages:

1. Defining the problem
2. Gathering the data
3. Organizing and analyzing the data
4. Forming a hypothesized cause
5. Verifying the hypothesis by tests, additional data, or calculations
6. Implementing a solution

The case histories in this book were best slotted into step 3 or 5; however, step 2 is also very important. Without valid field data, troubleshooting results will be suspect and are much like inadequate or faulty data when authorities analyze a crime. Improper and costly conclusions could be reached. Consider the case of a large steam turbine driving a centrifugal compressor. The compressor destroyed its internal labyrinth seals, and some of the rotor blades were damaged. There was debris in the bottom of the case that looked like chunks of carbon. The process computer history didn't show any obvious problems, nor did the unit's logs or maintenance files.

A lot of data are present, much of it obtained from the experienced unit operator and the machinist. The word *experienced* is so important! All too often a piece of equipment is "cleaned up" by the crew that disassembles the machine. Broken parts may be degreased, tossed about, wire brushed, or thrown away. This can destroy the evidence that is so valuable in determining the cause. The debris in the compressor was analyzed and found to be an elastomer product. The suction line was traced back to a lined valve that had deteriorated with age, with sections having broken loose.

Without the evidence that was collected, the machine would have been restarted after the repairs and probably would have failed again after more of the lining dislodged. Although the final solution to replace the valve was obvious, the cause was determined only by having the correct failure data and knowing what they meant.

REFERENCES

1. R. C. Juvinall, *Stress, Strain, and Strength*, McGraw-Hill, New York, 1967, p. 242.
2. *Calculation of Heater Tube Thickness in Petroleum Refineries*, API 530, American Petroleum Institute,
3. J. Miller and F. R. Lawson, Time, Temperature relationships for rupture creep stresses, *ASME Transactions*, Vol. 74, 1952.
4. R. J. Roark, *Formulas for Stress and Strength*, 4th ed., McGraw-Hill, New York, 1965.
5. R. E. Peterson, *Stress Concentration Design Factors*, Wiley, New York, 1958.
6. Agitator and Mixing Technology Seminar, Chemineer Inc., Dayton, OH.
7. *Bearing Technical Journal*, PT Components, Inc., 1982, p. A-25.
8. A. Blake, *Design of Mechanical Joints*, Marcel Dekker, New York, 1985, p. 123.
9. J. H. Bickford, *An Introduction to the Design and Behavior of Bolted Joints*, Marcel Dekker, New York, 1990, p. 606.
10. F. Kull, Stripping strength of tapped holes, *Product Engineering*, February 1957.
11. J. W. Hines, S. Jesse, A. Edmondson, and D. Nower, Study shows shaft misalignment reduces bearing life, *Maintenance Technology*, April 1999.
12. M. F. Spotts, *Design of Machine Elements*, 3rd ed., Prentice-Hall, Englewood Cliffs, NJ, 1965, p. 298.
13. H. Gartmann, *De Laval Engineering Handbook*, 3rd ed., McGraw-Hill, New York, 1970.
14. C. H. Jennings, Welding design, *ASME Transactions*, Vol. 58, p. 497.
15. *Recommended Practice: Fitness for Service*, API 579, American Petroleum Institute, January 2000.

16. H. P. Bloch and A. R. Budris, Life extension, in *Pump User's Handbook*, Fairmont Press, Lilburn, GA, 2004, p. 68.

17. D. J. Wulpi, *Understanding How Components Fail*, ASM Press, Washington, DC, 1990.

18. Deere & Company, *Identification of Parts Failures*, John Deere Publishing, Moline, IL, 1999.

19. SKF Industries, Inc., Bearing failures and their causes, presented at the Bearing Maintenance Seminar, 1985.

20. S. R. Lampman, ed., *Fatigue and Fracture*, Vol. 19, ASM Press, Washington, DC, 1996, p. 253.

21. *ASME Boiler and Pressure Vessel Code*, Section VIII, Division 2, ASME, New York, 1989.

22. J. C. Veiga, Teadit, in *Industrial Gaskets*, 2nd ed., Houston, TX, 1999, p. 18.

23. M. B. Peterson and W. O. Winer, eds., *Wear Control Handbook*, ASME, New York, 1980.

24. J. P. Den Hartog, *Mechanical Vibrations*, 4th ed., McGraw-Hill, New York, 1956.

25. E. J. Nestorides, ed., *A Handbook on Torsional Vibration*, B.I.C.E.R.A. Research Laboratory, Cambridge University Press, Cambridge, 1957.

26. A. Sofronas, Stopping torsional vibration, *Machine Design*, June 1974, p. 168.

27. R. D. Blevins, *Flow Induced Vibrations*, 2nd ed., Van Nostrand Reinhold, New York, 1990.

28. M. W. Kellogg Company, *Design of Piping Systems*, 2nd ed., Wiley, New York, 1956, p. 283.

29. Crane Engineering Division, *Flow of Fluids*, Technical Paper 410, Crane Company, New York, 1980.

30. I. J. Karassik, *Centrifugal Pump Clinic*, 2nd ed., Marcel Dekker, New York, 1989.

31. W. Drieder, Controlling centrifugal pumps, *Hydrocarbon Processing*, July 1995.

32. H. P. Bloch, ODR and DSS initiatives for reliability focused plants, *Hydrocarbon Processing*, May 2003.

33. I. H. Abbott and A. E. Von Doenhoff, *Theory of Wing Sections*, Dover, New York, 1959.

34. B. H. Jennings and S. R. Lewis, *Air Conditioning and Refrigeration*, International Textbook Company, Seranton, PA, 1965, p. 507.

35. J. H. Caldwell, *Preventative Maintenance: Reciprocating Gas Engines and Compressors*, Cooper Bessemer Industries.

36. J. Mancuso and J. Corcoran, What are the differences in high performance flexible couplings for turbomachinery, *Proceedings of the 32nd Turbomachinery Symposium*, Texas A&M University, College Station, TX, 2003.

37. J. R. Mancuso and R. Jones, Coupling interface connection, *Proceedings of the 30th Turbomachinery Symposium*, Texas A&M University, College Station, TX, 2001.

38. C. R. Hicks, *Fundamental Concepts in the Design of Experiments*, 2nd ed., Holt, Rinehart and Winston, New York, 1973.

39. J. K. Byers et al., *Pocket Handbook on Reliability*, USAAVSCOM Technical Report 77-16, U.S. Department of Commerce, Washington, DC, September 1975.

40. H. P. Bloch and F. K. Geitner, *Machinery Failure Analysis and Troubleshooting*, Gulf Publishing, Houston, TX.

41. R. B. Waterhouse, *Fretting Corrosion*, Pergamon Press, Elmsford, NY, 1972.

42. A. E. Anderson et al., *Friction, Lubrication, and Wear Technology*, Vol. 18, ASM Press, Washington, DC, 1992, p. 242.

43. R. S. Gill, Engineering Consultant, Houston, TX, personal communications, May 2004.

44. D. G. Sopwith, The distribution of load in screw-thread, *Institution of Mechanical Engineers, Proceedings*, Vol. 159, 1948.

45. S. Timoshenko, *Strength of Materials*, Part II, *Advanced Theory*, 3rd ed., Van Nostrand, Princeton, NJ, 1956, p. 383.

46. F. B. Seely and J. O. Smith, *Advanced Mechanics of Materials*, 2nd ed., Wiley, New York, 1965, p. 534.

47. *Rating the Pitting Resistance and Bending Strength of Spur and Helical Gear Teeth*, AGMA Standard Practice 218.01, American Gear Manufacturers Association, 1982.

48. C. Rauwendaal, *Polymer Extrusion*, Hanser Publishers, Munich, 1986, p. 352.

49. R. D. Carlile and R. T. Fenner, On the lubricating action of molten polymers in single screw extruders, *Journal of Mechanical Engineering Science*, Vol. 20, No. 2, 1978.

50. V. Anand and K. Das, Selecting screw base and hard facing materials, Part 1, *Plastics Machinery and Auxiliaries*, November 1993, p. 27.

51. J. P. Holman, *Heat Transfer*, McGraw-Hill, New York, 1963, p. 57.

52. J. R. Sims et al., FFS of thin areas, in *PVP*, Vol. 233, ASME, New York, 1992.

53. J. M. Barsom and S. T. Rolfe, *Fracture and Fatigue Control in Structures*, Prentice-Hall, Englewood Cliffs, NJ, 1987, p. 17.

54. P. M. Scott, T. L. Anderson, D. A. Osage and G. M. Wilkowski, *Review of Existing Fitness-for-Service Criteria for Crack-like Flaws*, Bulletin 430, WRC, New York, 1998.

55. *ASME Boiler and Pressure Vessel Code*, Section VIII, Division 1, ASME, New York, 1989.

56. A. Sofronas, Engineering case history series, Feb. 2001 through August 2005, *Hydrocarbon Processing Magazine*.

INDEX

Analytical Troubleshooting of Process Machinery and Pressure Vessels: Including Real-World Case Studies, by Anthony Sofronas
Copyright © 2006 John Wiley & Sons, Inc.